Imaginary Numbers

Imaginary Numbers

An Anthology of Marvelous Mathematical Stories, Diversions, Poems, and Musings

Edited by William Frucht

John Wiley & Sons, Inc.

New York | Chichester | Weinheim | Brisbane | Singapore | Toronto

This book is printed on acid-free paper. ♾

Permission acknowledgments can be found on pages 325–327.

Published by John Wiley & Sons, Inc.
Published simultaneously in Canada

ISBN 0-471-33244-5

Printed in the United States of America

10 9 8 7 6 5 4 3 2 1

Contents

Preface

In fifth grade I was cast as the Mock Turtle in my school's production of *Alice in Wonderland.* My mother was supposed to make my costume. Like many suburban matrons of the time, she owned a sewing machine but had only the vaguest idea how to use it. She managed to make me a light green hood and some little boot-like things for my feet (Mrs. Hackel, one of the teachers organizing the show, had to tell her to turn them inside out to hide the stitching), and the local A&S cheerfully sold her a pair of green tights and a green turtleneck shirt. But she was utterly defeated by the task of making a shell. So, just before the final dress rehearsal, after a whispered conversation with the director, Mrs. Hackel hurried off and returned fifteen minutes later with my turtle shell.

I found it distinctly disappointing: a flat, vaguely oval piece of corrugated brown cardboard, with livid green tortoise markings painted on one side, an elastic strap stapled on the back for hanging around one's neck, and down at the bottom a little brown tail, perhaps five inches long. It was not mentioned to me that this costume had a history.

The school, one of those sprawling one-story brick structures built in the late fifties to warehouse young baby boomers, was at the time less than ten years old. Some years before, when it was very new, the faculty had decided to introduce themselves to the local parents by putting on a show. My shell had been worn by a pompous and stupid assistant principal for some sort of soft-shoe routine. But the bottom of the shell, which came nearly to my knees, fell rather higher on him, and the charmingly curved tail looked like . . . Suffice it to say that for an audience used to seeing Ozzie and Harriet Nelson sleeping in twin beds, it caused a sensation.

The night of the show, as the curtain rose on my single scene, there was a discernible ripple of recognition before I'd spoken a line. Chuckles and titters soon gave way to general hilarity, and within minutes I had only to stand up to provoke gales of laughter. By the time I got to my big dance number, the Lobster Quadrille, they were rolling in the aisles. They gave me a standing ovation at the curtain call. Of course I knew nothing of the reason—I thought I had given the performance of my life.*

In another child, such an event might have engendered a lifelong love of the theater. Still others might have been frightened or even completely untouched. My response was different: I was moved to try to relive the experience, not of performing Lewis Carroll but of reading him. Much of my leisure time since then has been devoted to this pursuit.

I had read some classic children's literature—L. Frank Baum, Dr. Seuss, and the unbearably saccharine A. A. Milne were among my favorites—but all of it involved authors who had deliberately turned away from the moral complexities of the adult world in order to explore specific themes within the simplified realms of children. To a greater or lesser extent they all engaged the nostalgia of their adult readers. Having no need to be nostalgic for childhood (and being too naïve to fear or even much anticipate the transition to adulthood), I enjoyed these stories but was not deeply moved by them.

Carroll was something else entirely. He was no mere visitor in the worlds to which he carried Alice, he made no attempt to hide their unpleasant aspects or smooth their rough edges, and

*As things turned out, I had.

his creation of them seemed involuntary, even incontinent. I didn't feel he had "made up" these places, for me or anyone else, but that he had let me in. Wonderland and the Looking-Glass world were not merely whimsical but certifiably strange, and I took to them for the same reason boys take to heavy-metal music: for the sheer raw wildness of it. I had no idea that I was having my first encounter with a rigorously logical mind. Still, something of the obsessive mathematical edge to Carroll's thinking must have sunk in. It was about this time that I first began to sense the difference between humor and silliness—between the absurdly logical conclusion and one that is merely absurd.

I began to have tastes in reading, centered on the fantastic. Over the next several years I moved from Superman comics to the bad science fiction of people like Isaac Asimov and Robert Heinlein, to better science fiction (Ursula K. Le Guin, Cordwainer Smith, Olaf Stapledon), to contemporary writers such as Jorge Luis Borges and Gabriel García Marquez (as well as, inevitably, a lot of lesser ones). By the time I was midway through college, my reading life was thoroughly divided: realists like Fielding, Dos Passos, and Cather during the day, for class assignments; Gogol, Kafka, and Flannery O'Connor at night. These two strains became the sun and moon of my daily existence. Once I left school and had to work for a living, of course, I gave myself wholly over to the Dark Side.

The pieces collected here are just a tiny and very peculiar subset of that dreaming tradition. In a certain sense they represent my hopes for a truly literary science fiction—an SF connected to rather than estranged from the rest of literature. I want to be able to mention Connie Willis and Italo Calvino in the same breath: both project the geometry of spacetime onto human aspiration and suffering. Equally, I want to be able to mention Stanislaw Lem and Raymond Smullyan in the same breath; using mathematics to tell stories and using stories to explain mathematics are two sides of the same coin. They join what should never have separated: the scientist's and the artist's ways of uncovering truths about the world. These pieces represent an unconscious community of like interests, a tiny but steady current within a vast and growing stream.

While a small number of truly great writers, such as Calvino, Carroll, or Borges, wrote about mathematics quite often, in general surprisingly few stories or poems employ mathematical themes. This is surprising because, as John Paulos shows in *Once Upon a Number,* mathematics and narrative have a great deal in common. Both are abstract, symbolic ways of expressing our understanding of the world, even if the understandings they express seem almost completely incompatible. For a certain kind of writer, however, mathematical ideas are very compelling: their strange implications and inflexible logic lead to absurdities that make them seem at once perfectly rational and perfectly irrational, and this self-contradiction makes mathematics a perfect backdrop for the same self-contradiction in these writers' human characters. Many of the prose writers in this book gravitate toward fantasy or allegory because that is where this particular strain of irony can be best expressed.

A few mathematical themes crop up over and over: symmetry, infinity, probability, logic, the computer, the geometry of space-time. In most cases these themes are fairly obvious, but sometimes they are less so. Fritz Lieber's "Gonna Roll the Bones" centers on gambling with dice, which is one of the wellsprings of probability theory. But the hero's confidence to roll against the ultimate Big Gambler comes from his "almost incredible" skill at precision throwing: in the mine where he worked, "he sometimes amused himself by tossing little fragments of rock back into the holes from which they had fallen, so that they stuck there, perfectly fitted in, for at least a second"—a playful anticipation of what chaos theorists, a decade after Lieber's story, would famously call "sensitive dependence on initial conditions." "Giovanni and His Wife," by the Italian surrealist Tommaso Landolfi, relies implicitly on a certain mathematical understanding of music as a regular sequence of proportions and relationships (such as Douglas Hofstadter employs in *Gödel, Escher, Bach*), but also on the reader's intuition that when these relationships are not regular but random, an exact congruence of two such random sequences is improbable beyond the ability of mere numbers to express. Philip K. Dick's "The Golden Man" expresses a view of time drawn from David Bohm's "many worlds" interpretation of quantum mechanics. (This is the same view described in Borges's "The Garden of Forking Paths," one of several Borges stories that could easily have been included here.)

"More than iron, more than lead, more than gold I need electricity. . . . I need it for my dreams" becomes more compelling when one understands that the speaker, Racter, is a computer program. Everything it does is an electronic manipulation of abstract symbols according to mathematical instructions. In William Gibson's "Burning Chrome," that algorithmic world becomes a dark caricature—an image in black ice—of the evil and innocent passions of the real world.

J. G. Ballard's beautiful story "The Garden of Time," the least justifiable inclusion in the entire collection, is here only because it conveys to me some of the *feeling* of mathematical objects: the spare infinite plain; the mysterious garden full of strange transformations; the ambiguity of Alex and his wife's existence as both real entities and idealizations. If the "moddler" of Rudy Rucker's "A New Golden Age" were ever actually built, I hope that some especially poignant theorem would feel to me as this story does.

The spirit of Alice broods over this book. The paradox of reason is that its illumination shows us a universe of profound darkness: a vast place relentlessly driven by violent, inflexible laws—theoretically knowable yet of infinite depth—that care nothing for human desire or welfare. The paradox of unreason is that it is so fragile and self-contained—a bottle set adrift in a gale, without anchor or direction—and yet manages to survive. Alice, cast into Wonderland, has nothing to guide her through a logic-mad world except her absurd Victorian propriety and her ridiculous sense of courtesy. Yet when applied with sufficient faith and courage, these meager tools are somehow enough. The comforting light of familiar irrationality prevails over the darkness of reason.

But if this is a happy ending, it's at best an ironic one. Perhaps the final image presented by these stories and poems is of a child-like humanity capering innocently on the cosmic stage, unaware that the joke isn't even on us.

Oh time thy pyramids

—Jorge Luis Borges

The Form of Space

Italo Calvino

Translated by William Weaver

The equations of the gravitational field which
relate the curve of space to the distribution of
matter are already becoming common knowledge.

To fall in the void as I fell: none of you knows what that means. For you, to fall means to plunge perhaps from the twenty-sixth floor of a skyscraper, or from an airplane which breaks down in flight: to fall headlong, grope in the air a moment, and then the Earth is immediately there, and you get a big bump. But I'm talking about the time when there wasn't any Earth underneath or anything else solid, not even a celestial body in the distance capable of attracting you into its orbit. You simply fell, indefinitely, for an indefinite length of time. I went down into the void, to the most absolute bottom conceivable, and once there I saw that the extreme limit must have been much, much farther below, very remote, and I went on falling, to reach it. Since there were no reference points, I had no idea whether my fall was fast or slow. Now that I think about it, there weren't even any proofs that I was really falling: perhaps I had always remained immobile in the same place, or I was moving in an upward direction; since there was no above or below these were only nominal questions and so I might just as well go on thinking I was falling, as I was naturally led to think.

Assuming then that one was falling, everyone fell with the

same speed and rate of acceleration; in fact we were always more or less on the same level: I, Ursula H'x, Lieutenant Fenimore. I didn't take my eyes off Ursula H'x: she was very beautiful to see, and in falling she had an easy, relaxed attitude. I hoped I would be able sometimes to catch her eye, but as she fell, Ursula H'x was always intent on filing and polishing her nails or running her comb through her long, smooth hair, and she never glanced toward me. Nor toward Lieutenant Fenimore, I must say, though he did everything he could to attract her attention.

Once I caught him—he thought I couldn't see him—as he was making some signals to Ursula H'x: first he struck his two index fingers, outstretched, one against the other, then he made a rotating gesture with one hand, then he pointed down. I mean, he seemed to hint at an understanding with her, an appointment for later on, in some place down there, where they were to meet. All nonsense, I knew perfectly well: there were no meetings possible among us, because our falls were parallel and the same distance always remained between us. But the mere fact that Lieutenant Fenimore had got such ideas into his head—and tried to put them into the head of Ursula H'x—was enough to get on my nerves, even though she paid no attention to him, indeed she made a slight blurting sound with her lips, directed—I felt there was no doubt—at him. (Ursula H'x fell, revolving with lazy movements as if she were turning in her bed and it was hard to say whether her gestures were directed at someone else or whether she was playing for her own benefit, as was her habit.)

I too, naturally, dreamed only of meeting Ursula H'x, but since, in my fall, I was following a straight line absolutely parallel to the one she followed, it seemed inappropriate to reveal such an unattainable desire. Of course, if I chose to be an optimist, there was always the possibility that, if our two parallels continued to infinity, the moment would come when they would touch. This eventuality gave me some hope; indeed, it kept me in a state of constant excitement. I don't mind telling you I had dreamed so much of a meeting of our parallels, in great detail, that it was now a part of my experience, as if I had actually lived it. Everything would happen suddenly, with simplicity and naturalness: after the long separate journey, unable to move an inch closer to each other, after having felt her as an alien being for so long, a prisoner of her parallel route, then the consistency of space, instead of being impalpable as it had always been, would become more taut and, at the same time, looser, a condensing of the void which would seem to come not from outside but from within us, and would press me and Ursula H'x together (I had only to shut my eyes to see her come forward, in an attitude I recognized as hers even if it was different from all her habitual attitudes: her arms stretched down, along her sides, twisting her wrists as if she were stretching and at the same time writhing and leaning forward), and then the invisible line I was following would become a single line, occupied by a mingling of her and me where her soft and secret nature would be penetrated or rather would enfold and, I would say, almost absorb the part of myself that till then had been suffering at being alone and separate and barren.

Even the most beautiful dreams can suddenly turn into nightmares, and it then occurred to me that the meeting point of our two parallels might also be the point at which all parallels existing in space eventually meet, and so it would mark not only my meeting with Ursula H'x but also—dreadful prospect—a meeting with Lieutenant Fenimore. At the very moment when Ursula H'x would cease to be alien to me, another alien with his thin black mustache would share our intimacies in an inextricable way: this thought was enough to plunge me into the most tormented jealous hallucinations: I heard the cry that our meeting—hers and mine—tore from us melt in a spasmodically joyous unison and then—I was aghast at the presentiment—from that sound burst her piercing cry as she was violated—so, in my resentful bias, I imagined—from behind, and at the same time the Lieutenant's vulgar shout of triumph, but perhaps—and here my jealousy became delirium—these cries of theirs, hers and his—might also not be so different or so dissonant, they might also achieve a unison, be joined in a single cry of downright pleasure, distinct from the sobbing, desperate moan that would burst from my lips.

In this alternation of hopes and apprehensions I continued to fall, constantly peering into the depths of space to see if anything heralded an immediate or future change in our condition. A couple of times I managed to glimpse a universe, but it was far away and seemed very tiny, well off to the right or to the left; I barely had time to make out a certain number of galaxies like shining little dots collected into superimposed masses which revolved with a faint buzz, when everything would vanish as it had appeared, upwards or to one side, so that I began to suspect it had only been a momentary glare in my eyes.

"There! Look! There's a universe! Look over there! There's something!" I shouted to Ursula H'x, motioning in that direction; but, tongue between her teeth, she was busy caressing the smooth, taut skin of her legs, looking for those very rare and almost invisible excess hairs she could uproot with a sharp tug of her pincerlike nails, and the only sign she had heard my call might be the way she stretched one leg upwards, as if to exploit—you would have said—for her methodical inspection the dim light reflected from that distant firmament.

I don't have to tell you the contempt Lieutenant Fenimore displayed toward what I might have discovered on those occasions: he gave a shrug—shaking his epaulettes, his bandoleer, and the decorations with which he was pointlessly arrayed—and turned in the other direction, snickering. Unless he was the one (when he was sure I was looking elsewhere) who tried to arouse Ursula's curiosity (and then it was my turn to laugh, seeing that her only response was to revolve in a kind of somersault, turning her behind to him: a gesture no doubt disrespectful but lovely to see, so that, after rejoicing in my rival's humiliation, I caught myself envying him this, as a privilege), indicating a labile point fleeing through space, shouting: "There! There! A universe! This big! I saw it! It's a universe!"

I won't say he was lying: statements of that sort, as far as I know, were as likely to be true as false. It was a proved fact that, every now and then, we skirted a universe (or else a universe skirted us), but it wasn't clear whether these were a number of universes scattered through space or whether it was always the same universe we kept passing, revolving in a mysterious trajectory, or whether there was no universe at all and what we thought we saw was the mirage of a universe which perhaps had once existed and whose image continued to rebound from the walls of space like the rebounding of an echo. But it could also be that the universes had always been there, dense around us, and had no idea of moving, and we weren't moving, either, and everything was arrested forever, without time, in a darkness punctuated only by rapid flashes when something or someone managed for a moment to free himself from that sluggish timelessness and indicate the semblance of a movement.

All these hypotheses were equally worth considering, but they interested me only insofar as they concerned our fall and the possibility of touching Ursula H'x. In other words, nobody really knew anything. So why did that pompous Fenimore sometimes assume a superior manner, as if he were certain of things? He had realized that when he wanted to infuriate me the surest system was to pretend to a long-standing familiarity with Ursula H'x. At a certain point Ursula took to swaying as she came down, her knees together, shifting the weight of her body this way and that, as if wavering in an ever-broader zigzag: just to break the monotony of that endless fall. And the Lieutenant then also started swaying, trying to pick up her rhythm, as if he were following the same invisible track, or rather as if he were dancing to the sound of the same music, audible only to the two of them, which he even pretended to whistle, putting into it, on his own, a kind of unspoken understanding, as if alluding to a private joke among old boozing companions. It was all a bluff—I knew that, of course—but still it gave me the idea that a meeting between Ursula H'x and Lieutenant Fenimore might already have taken place, who knows how long ago, at the beginning of their trajectories, and this suspicion gnawed at me painfully, as if I had been the victim of an injustice. On reflecting, however, I reasoned that if Ursula and the Lieutenant had once occupied the same point in space, this meant that their respective lines of fall had since been moving apart and presumably were still moving apart. Now, in this slow but constant removal from the Lieutenant, it was more than likely that Ursula was coming closer to me; so the Lieutenant had little to boast of in his past conjunctions: I was the one at whom the future smiled.

The process of reasoning that led me to this conclusion was not enough to reassure me at heart: the possibility that Ursula H'x had already met the Lieutenant was in itself a wrong which, if it had been done to me, could no longer be redeemed. I must add that past and future were vague terms for me, and I couldn't make much distinction between them: my memory didn't extend beyond the interminable present of our parallel fall, and what might have been before, since it couldn't be remembered, belonged to the same imaginary world as the future, and was confounded with the future. So I could also suppose that if two parallels had ever set out from the same point, these were the lines that Ursula H'x and I were following (in this case it was nostalgia for a lost oneness that fed my eager desire to meet her); however, I was reluctant to believe in this hypothesis, because it might imply a progressive separation and perhaps her future arrival in the braid-festooned arms of Lieutenant Fenimore, but chiefly because I couldn't get out of the present except to imagine a different present, and none of the rest counted.

Perhaps this was the secret: to identify oneself so completely with one's own state of fall that one could realize the line followed in falling wasn't what it seemed but another, or rather to succeed in changing that line in the only way it could be changed, namely, by making it become what it had really always been. It wasn't through concentrating on myself that this idea came to me, though, but through observing, with my loving eye, how beautiful Ursula H'x was even when seen from behind, and noting, as we passed in sight of a very distant system of constellations, an arching of her back and a kind of twitch of her behind, but not so much the behind itself as an external sliding that seemed to rub past the behind and cause a not unpleasant reaction from the behind itself. This fleeting impression was enough to make me see our situation in a new way: if it was true that space with something inside is different from empty space because the matter causes a curving or a tautness which makes all the lines contained in space curve or tauten, then the line each of us was following was straight in the only way a straight line can be straight: namely, deformed to the extent that the limpid harmony of the general void is deformed by the clutter of matter, in other words, twisting all around this bump or pimple or excrescence which is the universe in the midst of space.

My point of reference was always Ursula and, in fact, a certain way she had of proceeding as if twisting could make more familiar the idea that our fall was like a winding and unwinding in a sort of spiral that tightened and then loosened. However, Ursula—if you watched her carefully—wound first in one direction, then in the other, so the pattern we were tracing was more complicated. The universe, therefore, had to be considered not a crude swelling placed there like a turnip, but as an angular, pointed figure where every dent or bulge or facet corresponded to other cavities and projections and notchings of space and of the lines we followed. This, however, was still a schematic image, as if we were dealing with a smooth-walled solid, a compenetration of polyhedrons, a cluster of crystals; in reality the space in which we moved was all battlemented and perforated, with spires and pinnacles which spread out on every side, with cupolas and balustrades and peristyles, with rose windows, with double- and triple-arched fenestrations, and while we felt we were plunging straight down, in reality we were racing along the edge of moldings and invisible friezes, like ants who, crossing a city, follow itineraries traced not on the street cobbles but along walls and ceilings and cornices and chandeliers. Now if I say city it amounts to suggesting figures that are, in some way, regular, with right angles and symmetrical proportions, whereas instead, we should always bear in mind how space breaks up around every cherry tree and every leaf of every bough that moves in the wind, and at every indentation of the edge of every leaf, and also it forms along every vein of the leaf, and on the network of veins inside the leaf, and on the piercings made every moment by the riddling arrows of light, all printed in negative in the dough of the void, so that there is nothing now that does not leave its print, every possible print of every possible thing, and together every transformation of these prints, instant by instant, so the pimple growing on a caliph's nose or the soap bubble resting on a laundress's bosom changes the general form of space in all its dimensions.

All I had to do was understand that space was made in this way and I realized there were certain soft cavities hollowed in it as welcoming as hammocks where I could lie joined with Ursula H'x, the two of us swaying together, biting each other in turn along all our persons. The properties of space, in fact, were such that one parallel went one way, and another in another way: I for example was plunging within a tortuous cavern while Ursula H'x was being sucked along a passage communicating with that same cavern so that we found ourselves rolling together on a lawn of algae in a kind of subspatial island, writhing, she and I, in every pose, upright and capsized, until all of a sudden our two straight lines resumed their distance, the same as always, and each continued on its own as if nothing had happened.

The grain of space was porous and broken with crevasses and dunes. If I looked carefully, I could observe when Lieutenant Fenimore's course passed through the bed of a narrow, winding canyon; then I placed myself on the top of a cliff and, at just the right moment, I hurled myself down on him, careful to strike him on the cervical vertebrae with my full weight. The bottom of such precipices in the void was stony as the bed of a dried-up stream, and Lieutenant Fenimore, sinking to the ground, remained with his head stuck between two spurs of rock; I pressed one knee into his stomach, but he meanwhile was crushing my knuckles against a cactus's thorns—or the back of a porcupine? (spikes, in any case, of the kind corresponding to certain sharp contractions of space)—to prevent me from grabbing the pistol I had kicked from his hand. I don't know how I happened, a moment later, to find myself with my head thrust into the stifling granulosity of the strata where space gives way, crumbling like sand; I spat, blinded and dazed; Fenimore had managed to collect his pistol; a bullet whistled past my ear, ricocheting off a proliferation of the void that rose in the shape of an anthill. And I fell upon him, my hands at his throat, to strangle him, but my hands slammed against each other with a "plop!": our paths had become parallel again, and Lieutenant Fenimore and I were descending, maintaining our customary distance, ostentatiously turning our backs on each other, like two people who pretend they have never met, haven't even seen each other before.

What you might consider straight, one-dimensional lines were similar, in effect, to lines of handwriting made on a white page by a pen that shifts words and fragments of sentences from one line to another, with insertions and cross-references, in the haste to finish an exposition which has gone through successive, approximate drafts, always unsatisfactory; and so we pursued each other, Lieutenant Fenimore and I, hiding behind the loops of the *l*'s, especially the *l*'s of the word "parallel," in order to shoot and take cover from the bullets and pretend to be dead and wait, say, till Fenimore went past in order to trip him up and drag him by his feet, slamming his chin against the bottoms of the *v*'s and the *u*'s and the *m*'s and the *n*'s which, written all evenly in an italic hand, became a bumpy succession of holes in the pavement (for example, in the expression "unmeasurable universe"), leaving him stretched out in a place all trampled with erasings and x-ings, then standing up there again, stained with clotted ink, to run toward Ursula H'x, who was trying to act sly, slipping behind the tails of the *f* which trail off until they become wisps, but I could seize her by the hair and bend her against a *d* or a *t* just as I write them now, in haste, bent, so you can recline against them, then we might dig a niche for ourselves down in a *g*, in the *g* of "big," a subterranean den which can be adapted as we choose to our dimensions, being made more cozy and almost invisible or else arranged more horizontally so you can stretch out in it. Whereas naturally the same lines, rather than remain series of letters and words, can easily be drawn out in their black thread and unwound in continuous, parallel, straight lines which mean nothing beyond themselves in their constant flow, never meeting, just as we never meet in our constant fall: I, Ursula H'x, Lieutenant Fenimore, and all the others.

A New Golden Age

Rudy Rucker

"It's like music," I repeated. Lady Vickers looked at me uncomprehendingly. Pale British features beneath wavy red hair, a long nose with a ripple in it.

"You can't hear mathematics," she stated. "It's just squiggles in some great dusty book." Everyone else around the small table was eating. White soup again.

I laid down my spoon. "Look at it this way. When I read a math paper it's no different than a musician reading a score. In each case the pleasure comes from the play of patterns, the harmonies and contrasts." The meat platter was going around the table now, and I speared a cutlet.

I salted it heavily and bit into the hot, greasy meat with pleasure. The food was second-rate, but it was free. The prospect of unemployment had done wonders for my appetite.

Mies van Koop joined the conversation. He had sparse curly hair and no chin. His head was like a large, thoughtful carrot with the point tucked into his tight collar. "It's a sound analogy, Fletch. But the musician can *play* his score, play it so that even a legislator . . ." he smiled and nodded donnishly to Lady Vickers. "Even a legislator can hear how beautiful Beethoven is."

"That's just what I was going to say," she added, wagging her finger at me. "I'm sure my husband has done lovely work, but the only way he knows how to show a person one of his beastly theorems is to make her swot through pages and pages of teeming little symbols."

Mies and I exchanged a look. Lord Vickers was a crank, an eccentric amateur whose work was devoid of serious mathematical interest. But it was thanks to him that Lady Vickers had bothered to come to our little conference. She was the only member of the Europarliament who had.

"Vat you think our chances are?" Rozzick asked her in the sudden silence, his mouth full of unchewed cauliflower.

"Dismal. Unless you can find some way of making your research appeal to the working man, you'll be cut out of next year's budget entirely. They need all the mathematics money for that new computer in Geneva, you know."

"We know," I said gloomily. "That's why we're holding this meeting. But it seems a little late for public relations. If only we hadn't let the government take over all the research funds."

"There's no point blaming the government," Lady Vickers said tartly. "People are simply tired of paying you mathematicians to make them feel stupid."

"Zo build the machine," Rozzick said with an emphatic bob of his bald little head.

"That's right," Mies said, "Build a machine that will play mathematics like music. Why not?"

Lady Vickers clapped her hands in delight and turned to me, "You mean you know how?"

Before I could say anything, Mies kicked me under the table. Hard. I got the message. "Well, we don't have quite all the bugs worked out . . ."

"But that's just too marvellous!" Lady Vickers gushed, pulling out a little appointment book. "Let's see . . . the vote on the math appropriation is June 4 . . . which gives us six weeks. Why don't you get your machine ready and bring it to Foxmire towards the end of May? The session is being held in London, you know, and I could bring the whole committee out to *feel* the beauty of mathematics."

I was having trouble moving my mouth. "Is planty time," Rozzick put in, his eyes twinkling.

Just then Watson caught the thread of the conversation. In the journals he was a famous mathematician . . . practically a grand old man. In conversation he was the callowest of eighteen-year-olds. "Who are you trying to kid, Fletcher?" He shook his head, and dandruff showered down on the narrow shoulders of his black suit. "There's no way . . ." He broke off with a yelp of pain. Mies was keeping busy.

"If you're going to make that train, we'd better get going," I said to Lady Vickers with a worried glance at my watch.

"My dear me, yes," she agreed, rising with me. "We'll expect you and your machine on May 23 then?" I nodded, steering her across the room. Watson had stuck his head under the table to see what was the matter. Something was preventing him from getting it back out.

When I got back from the train station, an excited knot of people had formed around Watson, Rozzick and Mies. Watson spotted me first, and in his shrill cracking voice called out, "Our pimp is here."

I smiled ingratiatingly and joined the group. "Watson thinks it's immoral to make mathematics a sensual experience," Mies explained. "The rest of us feel that greater exposure can only help our case."

"Where is machine?" Rozzick asked, grinning like a Tartar jack-o'-lantern.

"You know as well as I do that there is none. All I did was remark to Lady Vickers . . ."

"One must employ the direct stimulation of the brain," LaHaye put in. He was a delicate old Frenchman with a shock of luminous white hair.

I shook my head. "In the long run, maybe. But I can't quite see myself sticking needles in the committee's brain-stems five weeks from now. I'm afraid the impulses are going to have to come in through normal . . ."

"Absolute Film," Rozzick said suddenly. "Hans Richter and Oskar Fischinger invented in the 1920s. Abstract patterns on screen, repeating and differentiating. Is in Warszaw archives accessible."

"Derisory!" LaHaye protested. "If we make of mathematics

an exhibit, it should not be a tawdry *son et lumière*. Don't worry about needles, Dr. Fletcher. There are new field methods." He molded strange shapes in the air around his snowy head.

"He's right," Watson nodded. "The essential thing about mathematics is that it gives aesthetic pleasure without coming through the senses. They've already got food and television for their eyes and ears, their gobbling mouths and grubbing hands. If we're going to give them mathematics, let's sock it to them right in the old grey matter!"

Mies had taken out his pen and a pad of paper. "What type of manifold should we use as the parameter space?"

We couldn't have done it if we'd been anywhere else but the Center. Even with their staff and laboratories it took us a month of twenty-hour days to get our first working math player built. It looked like one of those domey old hair-dryers growing out of a file cabinet with dials. We called it a Moddler.

No one was very interested in being the first to get his brain mathed or modified or coddled or whatever. The others had done most of the actual work, so I had to volunteer.

Watson, LaHaye, Rozzick and Mies were all there when I snugged the Moddler's helmet down over my ears. I squeezed the switch on and let the electrical vortex fields swirl into my head.

We'd put together two tapes, one on Book I of Euclid's *Elements,* and the other on iterated ultrapowers of measurable cardinals. The idea was that the first tape would show people how to understand things they'd vaguely heard of . . . congruent triangles, parallel lines and the Pythagorean theorem. The second tape was supposed to show the power and beauty of flat-out pure mathematics. It was like we had two excursions: a leisurely drive around a famous ruin, and a jolting blast down a drag-strip out on the edge of town.

We'd put the first tape together in a sort of patchwork fashion, using direct brain recordings as well as artificially punched-in thought patterns. Rozzick had done most of this one. It was all visualized geometry: glowing triangles, blooming circles and the like. Sort of an internalized Absolute Film.

The final proof was lovely, but for me the most striking part was a series of food images which Rozzick had accidentally let slip into the proof that a triangle's area is one-half base times height.

"Since when are triangles covered with anchovy paste?" I asked Rozzick as Mies switched tapes.

"Is your vision clear?" LaHaye wanted to know. I looked around, blinking. Everything felt fine. I still had an afterglow of pleasure from the complex play of angles in Euclid's culminating proof that the square of the hypotenuse is equal to the sum of the squares on the two sides.

Then they switched on the second tape. Watson was the only one of us who had really mastered the Kunen paper on which this tape was based. But he'd refused to have his brain patterns taped. Instead he'd constructed the whole thing as an artificial design in our parameter space.

The tape played in my head without words or pictures. There was a measurable cardinal. Suddenly I knew its properties in the same unspoken way that I knew my own body. I did something to the cardinal and it transformed itself, changing the concepts clustered around it. This happened over and over. With a feeling of light-headedness, I felt myself moving outside of this endless self-transformation . . . comprehending it from the outside. I picked out a certain subconstellation of the whole process, and swathed it in its logical hull. Suddenly I understood a theorem I had always wondered about.

When the tape ended I begged my colleagues for an hour of privacy. I had to think about iterated ultrapowers some more. I rushed to the library and got out Kunen's paper. But the lucidity was gone. I started to stumble over the notation, the subscripts and superscripts; I was stumped by the gappy proofs; I kept forgetting the definitions. Already the actual content of the main theorem eluded me. I realized then that the Moddler was a success. You could *enjoy* mathematics—even the mathematics you couldn't normally *understand*.

We all got a little drunk that night. Somewhere towards midnight I found myself walking along the edge of the woods with Mies. He was humming softly, beating time with gentle nods of his head.

We stopped while I lit my thirtieth cigarette of the day. In the match's flare I thought I caught something odd in Mies's expression. "What is it?" I asked, exhaling smoke.

"The music . . ." he began. "The music most people listen to is not good."

I didn't see what he was getting at, and started my usual defense of rock music.

"Muzak," Mies interrupted. "Isn't that what you call it . . . what they play in airports?"

"Yeah. Easy listening."

"Do you really expect that the official taste in mathematics will be any better? If everyone were to sit under the Moddler . . . what kind of mathematics would they ask for?"

I shrank from his suggestion. "Don't worry, Mies. There are objective standards of mathematical truth. No one will undermine them. We're headed for a new golden age."

LaHaye and I took the Moddler to Foxmire the next week. It was a big estate, with a hog wallow and three holes of golf between the gatehouse and the mansion. We found Lord Vickers at work on the terrace behind his house. He was thick-set and sported pop-eyes set into a high forehead.

"Fletcher and LaHaye," he exclaimed. "I am honored. You arrive opportunely. Behold." He pulled a sheet of paper out of his special typewriter and handed it to me.

LaHaye was looking over my shoulder. There wasn't much to see. Vickers used his own special mathematical notation. "It would make a nice wallpaper," LaHaye chuckled, then added quickly, "Perhaps if you once explained the symbolism . . ."

Lord Vickers took the paper back with a hollow laugh. "You know very well that my symbols are all defined in my *Thematics and Metathematics* . . . a book whose acceptance you have tirelessly conspired against."

"Let's not open old wounds," I broke in. "Dr. LaHaye's

remark was not seriously intended. But it illustrates a problem which every mathematician faces. The problem of communicating his work to non-specialists, to mathematical illiterates." I went on to describe the Moddler while LaHaye left to supervise its installation in Lord Vickers' study.

"But this is fantastic," Vickers exclaimed, pacing back and forth excitedly. A large Yorkshire hog had ambled up to the edge of the terrace. I threw it an apple.

Suddenly Vickers was saying, "We must make a tape of *Thematics and Metathematics*, Dr. Fletcher." The request caught me off-guard.

Vickers had printed his book privately, and had sent a copy to every mathematician in the world. I didn't know of anyone who had read it. The problem was that Vickers claimed he could do things like trisect angles with ruler and compass, give an internal consistency proof for mathematics, and so on. But we mathematicians have rigorous proofs that such things are impossible. So we knew in advance that Vickers' work contained errors, as surely as if he had claimed to have proved that he was twenty meters tall. To master his eccentric notation just to find out his specific mistakes seemed no more worthwhile than looking for the leak in a sunken ship.

But Lord Vickers had money and he had influence. I was glad LaHaye wasn't there to hear me answer, "Of course, I'd be glad to put it on tape."

And, God help me, I did. We had four days before Lady Vickers would bring the Appropriations Committee out for our demonstration. I spent all my waking time in Vickers' study, smoking his cigarettes and punching in *Thematics and Metathematics.*

It would be nice if I could say I discovered great truths in the book, but that's not the way it was. Vickers' work was garbage, full of logical errors and needless obfuscation. I refrained from trying to fix up his mistakes, and just programmed in the patterns as they came. LaHaye flipped when he found out what I was up to. "We have prepared a feast for the mind," he complained, "And you foul the table with this, this . . ."

"Think of it as a ripe Camembert," I sighed. "And serve it last. They'll just laugh it off."

Lady Vickers was radiant when she heard I'd taped *Thematics and Metathematics.* I suggested that it was perhaps too important a work to waste on the Appropriations Committee, but she wouldn't hear of passing it up.

Counting her, there were five people on the committee. LaHaye was the one who knew how to run the Moddler, so I took a walk while he ran each of the legislators through the three tapes.

It was a hot day. I spotted some of those hogs lying on the smooth hard earth under a huge beech tree, and I wandered over to look at them. The big fellow I'd given the apple was there, and he cocked a hopeful eye at me. I spread out my empty hands, then leaned over to scratch his ears. It was peaceful with the pigs, and after a while I lay down and rested my head on my friend's stomach. Through the fresh green beech leaves I could see the taut blue sky.

Lady Vickers called me in. The committee was sitting around the study working on a couple of bottles of Amontillado. Lord Vickers was at the sideboard, his back turned to me. LaHaye looked flushed and desperate.

"Well," I said.

"They didn't like the first tape . . ." LaHaye began.

"Dreary, dreary," Lady Vickers cried.

"We are not schoolchildren," another committee member put in.

I felt the floor sinking below me. "And the second tape?"

"I don't see how you can call that mathematics," Lady Vickers declaimed.

"There were no equations," someone complained.

"And it made me dizzy," another added.

"Here's to the new golden age of mathematics," Lord Vickers cried suddenly.

"To *Thematics and Metathematics,*" his wife added, lifting her glass. There was a chorus of approving remarks.

"That was the real thing."
"Plenty of logic."
"And so many symbols!"
Lord Vickers was smiling at me from across the room.
"There'll be a place for you at my new institute, Fletcher."
I took a glass of sherry.

A Serpent with Corners

Lewis Carroll

(from *A Tangled Tale*)

Water, water, everywhere,
Nor any drop to drink.

"It'll just take one more pebble."

"Whatever *are* you doing with those buckets?"

The speakers were Hugh and Lambert. Place, the beach of Little Mendip. Time 1:30 P.M. Hugh was floating a bucket in another a size larger, and trying how many pebbles it would carry without sinking. Lambert was lying on his back, doing nothing.

For the next minute or two Hugh was silent, evidently deep in thought. Suddenly he started. "I say, look here, Lambert!" he cried.

"If it's alive, and slimy, and with legs, I don't care to," said Lambert.

"Didn't Balbus say this morning that, if a body is immersed in liquid it displaces as much liquid as is equal to its own bulk?" said Hugh.

"He said things of that sort," Lambert vaguely replied.

"Well, just look here a minute. Here's the little bucket almost quite immersed: so the water displaced ought to be just about the same bulk. And now just look at it!" He took out the little bucket as he spoke, and handed the big one to Lambert. "Why, there's hardly a teacupful! Do you mean to say *that* water is the same bulk as the little bucket?"

"Course it is," said Lambert.

"Well, look here again!" cried Hugh, triumphantly, as he poured the water from the big bucket into the little one. "Why, it doesn't half fill it!"

"That's *its* business," said Lambert. "If Balbus says it's the same bulk, why, it *is* the same bulk, you know."

"Well, I don't believe it," said Hugh.

"You needn't," said Lambert. "Besides, it's dinnertime. Come along."

They found Balbus waiting dinner for them, and to him Hugh at once propounded his difficulty.

"Let's get you helped first," said Balbus, briskly cutting away at the joint. "You know the old proverb, 'Mutton first, mechanics afterwards'?"

The boys did *not* know the proverb, but they accepted it in perfect good faith, as they did every piece of information, however startling, that came from so infallible an authority as their tutor. They ate on steadily in silence and, when dinner was over, Hugh set out the usual array of pens, ink, and paper, while Balbus repeated to them the problem he had prepared for their afternoon's task.

"A friend of mine has a flower-garden—a very pretty one, though no great size—"

"How big is it?" said Hugh.

"That's what *you* have to find out!" Balbus gayly replied. "All *I* tell you is that it is oblong in shape—just half a yard longer than its width—and that a gravelwalk, one yard wide, begins at one corner and runs all round it."

"Joining into itself?" said Hugh.

"*Not* joining into itself, young man. Just before doing *that,* it turns a corner, and runs round the garden again, alongside of the first portion, and then inside that again, winding in and in, and each lap touching the last one, till it has used up the whole of the area."

"Like a serpent with corners?" said Lambert.

"Exactly so. And if you walk the whole length of it, to the last inch, keeping in the centre of the path, it's exactly two miles and half a furlong. Now, while you find out the length and breadth of the garden, I'll see if I can think out that sea-water puzzle."

"You said it was a flower-garden?" Hugh inquired, as Balbus was leaving the room.

"I did," said Balbus.

"Where do the flowers grow?" said Hugh. But Balbus thought it best not to hear the question. He left the boys to their problem, and, in the silence of his own room, set himself to unravel Hugh's mechanical paradox.

"To fix our thoughts," he murmured to himself, as, with hands deep-buried in his pockets, he paced up and down the room, "we will take a cylindrical glass jar, with a scale of inches marked up the side, and fill it with water up to the 10-inch mark: and we will assume that every inch depth of jar contains a pint of water. We will now take a solid cylinder, such that every inch of it is equal in bulk to *half* a pint of water, and plunge 4 inches of it into the water, so that the end of the cylinder comes down to the 6-inch mark. Well, that displaces 2 pints of water. What becomes of them? Why, if there were no more cylinder, they would lie comfortably on the top, and fill the jar up to the 12-inch mark. But unfortunately there *is* more cylinder, occupying half the space between the 10-inch and the 12-inch marks, so that only *one* pint of water can be accommodated there. What becomes of the other pint? Why, if there were no more cylinder, it would lie on the top, and fill the jar up to the 13-inch mark. But unfortunately—Shade of Newton!" he exclaimed, in sudden accents of terror. "When *does* the water stop rising?"

A bright idea struck him. "I'll write a little essay on it," he said.

BALBUS'S ESSAY

"When a solid is immersed in a liquid, it is well known that it displaces a portion of the liquid equal to itself in bulk, and that the level of the liquid rises just so much as it would rise if a quantity of liquid had been added to it, equal in bulk to the solid. Lardner says precisely the same process occurs when a solid is *partially* immersed: the quantity of liquid displaced, in this case, equalling the portion of the solid which is immersed, and the rise of the level being in proportion.

"Suppose a solid held above the surface of a liquid and partially immersed: a portion of the liquid is displaced, and the level of the liquid rises. But, by this rise of level, a little bit more of the solid is of course immersed, and so there is a new displacement of a second portion of the liquid, and a consequent rise of level. Again, this second rise of level causes a yet further immersion, and by consequence another displacement of liquid and another rise. It is self-evident that this process must continue till the entire solid is immersed, and that the liquid will then begin to immerse whatever holds the solid, which, being connected with it, must for the time be considered a part of it. If you hold a stick, six feet long, with its ends in a tumbler of water, and wait long enough, you must eventually be immersed. The question as to the source from which the water is supplied—which belongs to a high branch of mathematics, and is therefore beyond our present scope—does not apply to the sea. Let us therefore take the familiar instance of a man standing at the edge of the sea, at ebb-tide, with a solid in his hand, which he partially immerses: he remains steadfast and unmoved, and we all know that he must be drowned. The multitudes who daily perish in this manner to attest a philosophical truth, and whose bodies the unreasoning wave casts sullenly upon our thankless shores, have a truer claim to be called the martyrs of science than a Galileo or a Kepler. To use Kossuth's eloquent phrase, they are the unnamed demigods of the nineteenth century."*

* * *

* *Note by the writer.*—For the above essay I am indebted to a dear friend, now deceased.

"There's a fallacy *somewhere,*" he murmured drowsily, as he stretched his long legs upon the sofa. "I must think it over again." He closed his eyes, in order to concentrate his attention more perfectly, and for the next hour or so his slow and regular breathing bore witness to the careful deliberation with which he was investigating this new and perplexing view of the subject.

A Positive Reminder

J. A. Lindon

A carpenter named Charlie Bratticks,
Who had a taste for mathematics,
One summer Tuesday, just for fun,
Made a wooden cube side minus one.

Though this to you may well seem wrong,
He made it *minus* one foot long,
Which meant (I hope your brains aren't
frothing)
Its length was one foot less than nothing.

In width the same (you're not asleep?)
And likewise minus one foot deep;
Giving, when multiplied (be solemn!).
Minus one cubic foot of volume.

With sweating brow this cube he sawed
Through areas of solid board;
For though each cut had minus length,
Minus *times* minus sapped his strength.

A second cube he made, but thus:
This time each one foot length was plus:
Meaning of course that here one put
For volume: *plus* one cubic foot.

So now he had, just for his sins,
Two cubes as like as deviant twins:
And feeling one should know the worst,
He placed the second in the first.

One plus, one minus—there's no doubt
The edges simply cancelled out;
So did the volume, nothing gained;
Only the surfaces remained.

Well may you open wide your eyes,
For these were now of double size,
On something which, thanks to his skill,
Took up no room and measured nil.

From solid ebony he'd cut
These bulky cubic objects, but
All that remained was now a thin
Black sharply-angled sort of skin

Of twelve square feet—which though not
small,
Weighed nothing, filled no space at all,
It stands there yet on Charlie's floor;
He can't think what to use it for!

How Kazir Won His Wife

Raymond Smullyan

Annabelle and Alexander were tired after their long, winding climb to the Sorcerer's castle. But the Sorcerer, whom they discovered to be a most delightful and hospitable chap, served them a delicious brew consisting of equal parts of Ceylon tea and Chinese cinnamon wine, and they were instantly revived.

"Tell me," said Annabelle, whose interests were generally quite practical, "how did you ever establish your reputation here as a Sorcerer?"

"Ah, that's an amusing tale," said the Sorcerer, rubbing his hands. "I started my career here just twelve years ago. But I really owe my beginnings to the philosopher Nelson Goodman, who taught me a clever logical trick some forty years ago."

"What's the trick?" asked Alexander.

"Do you realize," said the Sorcerer, "that despite the fact that every inhabitant of this island is either a knight who only tells the truth or a knave who only tells lies, you can discover the truth or falsity of any proposition simply by asking any inhabitant only one question? And the odd part is that after he answers, you won't know whether his answer is the truth or a lie."

"That does sound clever," said Alexander.

"Well, that's how I got my position here," said the Sorcerer with a laugh. "You see, the first day I set foot on this island, I decided to go to the King's palace and apply for the job of sorcerer. The problem was I didn't know just where the palace was. At one point I came to a fork in the road; I knew that one of the two branches led to the palace, but I didn't know which one. There was a native standing at the fork; surely *he* would know which road was correct, but I didn't know whether he was a knight or a knave. Nevertheless, using the Goodman principle, I was able to find the correct road by asking him only one question answerable by yes or no."

What question did the Sorcerer ask him?

If you ask the native whether the left road is the one that leads to the palace, the question will be useless, since you have no idea whether he is a knight or a knave. The right question to ask is "Are you the type who would claim *that the left road leads to the palace?" After getting an answer, you will have no idea whether he is a liar or a truth-teller, but you* will *know which road to take. More specifically, if he answers yes, you should take the left road; if he answers no, you should take the right road. The proof of this is as follows.*

Suppose he answers yes. If he is a knight, then he has told the truth; he is the type who would claim that the left road leads to the palace. Hence the left road is the road to take. On the other hand, if he is a knave, his answer is a lie, which means he is not *the type who would claim that the left road leads to the palace; only one of the opposite type—a knight—would claim that the left road leads to the palace. But since a knight would claim that the left road leads to the palace, then again the left road really does lead to the palace. Regardless of whether the yes answer is the truth or a lie, the left road leads to the palace.*

Suppose now that the native answers no. If he is truthful, then he is really not the type who would claim that the left road leads to the palace; only a knave would claim that. And since a knave would make that claim, the claim is false, which means that the left road does not *lead to the palace. On the other hand if he is lying, then he really* would *claim that the left road leads to the palace (since he says he wouldn't), but, being a knave, his claim would be false, which means that the left road* doesn't *lead to the palace. This proves that if he answers no, the right road is the one to take, regardless of whether the native lied or told the truth.*

The Goodman principle is really a remarkable thing. Using it, one can extract any information from one who either always lies or always tells the truth. Of course, this strategy won't work on someone who sometimes tells the truth and sometimes lies.

"And so," said the Sorcerer, "I found the correct road and took it. I got lost a few more times along the way, but even that was for the best. The native I asked was so astounded at what I had done that he immediately told many of the island's other inhabitants, and the news reached the King before I did. The King was absolutely delighted with the incident and hired me on the spot! I've been doing good business ever since.

"And now," continued the Sorcerer, as he removed a beautifully bound volume from a shelf, "I have here an exceedingly rare and curious old book, written in Arabic, known as the *Tellmenow Isitsöornot.* I learned of it from the American author Edgar Allan Poe, in his story entitled 'The Thousand-and-Second Tale of Scheherazade.' Poe found this odd tale of Scheherazade in the *Isitsöornot*, but there are many other remarkable tales that he never mentioned and that are still comparatively unknown. For example, have you heard of the 'Five Tales of Kazir'?"

The two guests shook their heads.

"I thought not. These stories are of particular interest to logicians. Here, I will translate them for you.

"Once there was a young man named Kazir whose main ambition in life was to marry a king's daughter. He obtained an audience with the king of his country and frankly confessed his desire.

" 'You seem like a personable young man,' said the King, 'and I am sure my unmarried daughter will like you. But first you must pass a test. I happen to have two daughters named Amelia and Leila; one is married and the other is not. If you can pass the test, and if my unmarried daughter approves of you, then you may marry her.'

" 'Is Amelia or Leila the married one?' asked the suitor.

" 'Ah, that is for you to find out,' replied the King. 'That is your test.'

" 'Let me explain further,' he continued. 'My two daughters

are identical twins, yet temperamentally poles apart: Leila always lies and Amelia always tells the truth.'

" 'How extraordinary!' exclaimed the suitor.

" 'Most extraordinary, indeed,' replied the King. 'They have been like this from early childhood. Anyway, when I strike the gong, both daughters will appear. Your task is to determine whether Amelia or Leila is the married one. Of course you will not be told which is Amelia and which is Leila, nor will you be told which of them is married. You are allowed to ask just one question to just one of the two; then you must deduce the name of my married daughter.' "

"Oh, I get it," interrupted Annabelle. "Kazir used the Goodman principle. Is that it?"

"No," replied the Sorcerer. "The suitor happened to know the Goodman principle—though not by that name, of course. His tutor, a venerable old dervish, had taught it to him years before. And so Kazir was overjoyed and thought, 'All I need do is ask either daughter whether she is the type who would claim that Amelia is married. If she answers yes, then Amelia is married; if she answers no, then Leila is married. It's as simple as that.'

"But it *wasn't* as simple as that," continued the Sorcerer. "It so happened that the King could tell from the suitor's triumphant expression that he knew the Goodman principle. So the King said, 'I know what you're thinking, but I'm not going to let you use that logical trick on me. If your question contains more than three words, I'll have you executed on the spot!'

" 'Only three words?' cried Kazir.

" 'Only three words,' repeated the King.

"The gong was struck, and the two daughters appeared," What three-word question should the suitor ask to determine the name of the King's married daughter?

The suitor should ask, "Are you married?"

Suppose the daughter to whom he puts the question answers yes. She is either Amelia or Leila, but we don't know which. Suppose she is Amelia. Then her answer is truthful; Amelia really is married. But suppose she is Leila. Then her answer is a lie; Leila is not mar-

ried, so it must be Amelia who is married. Regardless of whether she is Amelia or Leila, if she answers yes, then Amelia must be married.

Suppose, on the other hand, that the daughter answers no. If she is Amelia, then her answer is truthful, which means that Amelia is not married; hence Leila is married. On the other hand, if Leila answers, then her answer is a lie, which means that she, Leila, is married. Regardless of whether the answer no is true, Leila is the married daughter.

"Did the suitor pass the test?" Annabelle asked.

"Alas, no," replied the Sorcerer. "If he hadn't been restricted to a three-word question, he would have had no trouble at all. But, confronted with a totally new situation, he was completely flustered. He just stood there, unable to utter a word."

"So what happened?" asked Alexander.

"He was dismissed from court. But then a curious thing happened. The unmarried daughter had taken a liking to Kazir, and pleaded with her father to summon him back the next day to take another test. Somewhat reluctantly, the King agreed.

"The next day, when the suitor entered the throne room, he exclaimed, 'I've thought of the right question.'

" 'Too late now,' said the King. 'You'll have to take another test. When I strike the gong, again my two daughters will appear (veiled, of course). One will be dressed in blue and the other in green. Your task now is *not* to find out the *name* of my married daughter, but which of the two—the one in blue or the one in green—is *unmarried*. Again you may ask only one question, and the question may not contain more than three words.'

"The gong was struck, and the two daughters appeared."

What three-word question should the suitor ask this time?

The question "Are you married?" will be of no help in this situation; the question to ask is "Is Amelia married?" If the daughter addressed answers yes, then she is married, regardless of whether she lies or tells the truth; if she answers no, then she is not married.

Suppose she answers yes. If she is Amelia, then her answer is truthful; Amelia is *married. On the other hand, if she is Leila, then her answer is a lie; Amelia is not married, so Leila is married. Thus, a yes answer indicates that the daughter who answers is married. (I leave it to the reader to verify that a no answer indicates that the person addressed is not married.)*

Of course, the question "Is Leila married?" would serve equally well; a yes answer would then indicate that the daughter who answers is not *married, a no answer that she is.*

As the Sorcerer explained to Annabelle and Alexander, there is a pretty symmetry between this problem and the last: to find out if the daughter addressed is married, you ask, "Is Amelia married?"; whereas if you wish to find out whether Amelia is married, you ask, "Are you *married?" These two questions have the curious property that asking either one will enable you to deduce the correct answer to the other.*

"Again the suitor failed," continued the Sorcerer, "but the unmarried daughter was more fond of him than ever. The King could not resist her pleadings, and so he agreed to give Kazir a third test the next day.

" 'This time,' said the King to the suitor, 'when I strike the gong, just one daughter will appear. Your task now is to find out her first name. Again you may ask only one question, and it may contain no more than three words.' "

What question should the suitor ask?

This problem is simpler than the preceding two. All the suitor need ask is any three-word question whose answer he already knows, such as "Does Leila lie?" To this particular question, Amelia would obviously answer yes and Leila would answer no.

" 'Now really,' said the King to his daughter after the suitor failed the third test, 'are you sure you want to marry him? He strikes me as quite a simpleton. The last test was ridiculously easy, and you know it.'

" 'He was just nervous,' responded the daughter. 'Please try him once more.'

"Well," said the Sorcerer, "for the next test the suitor was told that after the gong was struck the unmarried daughter would appear. The suitor was to ask her just one three-word question answerable by yes or no. If she answered yes, the suitor could marry her. If she answered no, he could not."

What question should the suitor ask?

The question "Are you Amelia?" would work perfectly well. Amelia, who is truthful, would answer yes, and Leila, who lies, would also answer yes—that is, she would falsely claim to be Amelia.

" 'This is trying my patience,' said the King to his daughter after the suitor had failed a fourth time. 'No more tests!'

" 'Just one more,' begged the daughter. 'I promise it will be the last.'

" 'All right, the very last one, you understand?'

"The daughter promised not to ask for any more tests, and so the King agreed.

" 'Now then,' said the King quite sternly to the suitor (who was trembling like a leaf), 'you have already failed four tests. You seem to have difficulty with these three-word questions, and so I will relax that requirement.'

" 'What a relief!' thought the suitor.

" 'When I sound the gong,' said the King, 'again only one daughter will appear; she may be the married one or the unmarried one. You are to ask her only one question answerable by yes or no, but the question may contain as many words as you like. From her answer, you must deduce *both* her name *and* whether she is married.' "

Can you think of a question that will work?

The king (evidently sick by now of the whole business) has given the suitor an impossible task. In each of the first four tests, the suitor had to determine which of two possibilities held. In this test, however, he is to determine which of four possibilities holds. (The four possibilities are: 1. the daughter addressed is Amelia and married; 2. she is Amelia and unmarried; 3. she is Leila and married; 4. she is Leila and unmarried.) However, there are only two possible responses to the suitor's question (either yes or no, since the question is required to have a yes/no answer). And with only two possible responses, it is impossible to determine which of four possibilities holds.

When I say that the task is impossible, I merely mean that there is no yes/no question that is bound *to work. There are several possible questions (at least four) that* might *happen to work if the suitor is lucky. For example, consider the question "Is it the case that you are married and that your name is Amelia?" Leila would answer yes to this (regardless of whether she is married because a compound statement—one with two parts joined by "and"—is false if* both *parts or only* one *part of the statement is false); Amelia, if she is married, would answer yes and, if she is unmarried, would answer no. So a yes answer would leave the suitor in the dark, whereas a no answer would indicate for sure that the one addressed is Amelia and unmarried. So if the suitor asked that question, he would have a twenty-five percent chance of finding out which one of the four possibilities actually holds. But no single question could ensure a* certainty *of finding out which one of the four possibilities holds.*

"So they never did get married?" asked Annabelle.

"The King never consented," replied the Sorcerer. "But the daughter was so furious at the unfairness of the last question that she felt perfectly justified in marrying Kazir without her father's permission. The two eloped and, according to the *Isitsöornot,* lived happily ever after.

"There is a valuable lesson to be learned from all this," continued the Sorcerer. "Never rely too much on general principles and routine mechanical methods. Certain classes of problems can be solved by such methods, but they completely lose their interest once the general principle is discovered. It is, of course, good to know these general principles—indeed, science and mathematics couldn't advance without them. But to depend on principles while neglecting intuition is a shame. The King was very clever to construct his tests so that the mere knowledge of a general principle—in this case the Goodman principle—would be of no avail. Each of his tests required a bit of ingenuity. Here the suitor failed, because of a lack of creative thinking."

"I would like to know one more thing," said Annabelle. "Is it recorded which of the two daughters Kazir married?"

"Oh, yes," replied the Sorcerer. "Fortunately for the suitor it was Amelia, the truthful one. This, I believe, contributed to their living happily ever after.

"The story of Leila's marriage," he continued, "was also recorded in the *Isitsöornot*, and I find this story particularly droll. It seems that Leila detested her suitor, but when he asked her one day, 'Would you like to marry me?', she, being a perpetual liar, said yes. And so they were married.

"So you see," concluded the Sorcerer, "perpetual lying sometimes has its dangers!"

Well, they all had a hearty laugh over that story. Then Annabelle and Alexander rose and thanked the Sorcerer for a most entertaining and instructive afternoon, explaining that they had to get back and make preparations for their departure the next day.

"Have you heard about the epidemic that hit this island three years ago?" asked the Sorcerer.

They shook their heads.

"Oh, I would like to tell you all about it, but unfortunately I can't today. I'm expecting a visit from the island Astrologer any minute. Why don't you stay a few more days?"

"My parents will be worried about me," explained Annabelle.

"Not so," replied the Sorcerer. "I've already dispatched a message to your island explaining that you are here and well. Your father knows that I am a knight."

This reassured the couple, and they agreed to visit the Sorcerer the next day.

"Who is this Astrologer?" asked Alexander, as they were leaving.

"Oh, he's a complete imbecile, as well as a knave and a charlatan. I have to put up with him; it's a matter of island politics. But I advise you to have nothing to do with him."

Just then, the Astrologer entered the room, nodded to the

couple (whom he had never seen before), and said to the Sorcerer, "She is the Queen of Sheba and he is King Solomon, you know."

"Typical," said the Sorcerer to the departing couple with a knowing wink.

11 May 1905

Alan Lightman

(from *Einstein's Dreams*)

Reminiscent of Italo Calvino's *Invisible Cities, Einstein's Dreams* centers on the young Albert Einstein, employed at the Patent Office in Berne, Switzerland, during his *annus mirabilis* of 1905, when he wrote three papers that forever altered modern physics. In one of these, Einstein presented his Special Theory of Relativity, which does away with the idea of an absolute time and proposes instead that the passage of time and even the order of events depend on how fast the observer is moving relative to those events. What this theory implies about the true nature of time—or even its existence—has barely begun to be explored.

As the historian of science Arthur I. Miller has written, "Einstein believed that scientific theories are a means for extending our intuition beyond our sense experiences." Alan Lightman imagines Einstein trying to imagine time—to explore with his intuition the realms made possible by his mathematics. Each of thirty "dreams" presents a different conception of time, as enacted by the mundane and picturesque yet somehow self-aware city of Berne.

On 11 May 1905 Einstein dreams about entropy, the tendency toward increasing disorder that is the only reason known to physics for time to move in one direction rather than another.

Walking on the Marktgasse, one sees a wondrous sight. The cherries in the fruit stalls sit aligned in rows, the hats in the millinery shop are neatly stacked, the flowers on the balconies are arranged in perfect symmetries, no crumbs lie on the bakery floor, no milk is spilled on the cobblestones of the buttery. No thing is out of place.

When a gay party leaves a restaurant, the tables are more tidy than before. When a wind blows gently through the street, the street is swept clean, the dirt and dust transported to the edge of town. When waves of water splash against the shore, the shore rebuilds itself. When leaves fall from the trees, the leaves line up like birds in V-formation. When clouds form faces, the faces stay. When a pipe lets smoke into a room, the soot drifts toward a corner of the room, leaving clear air. Painted balconies exposed to wind and rain become brighter in time. The sound of thunder makes a broken vase reform itself, makes the fractured shards leap up to the precise positions where they fit and bind. The fragrant odor of a passing cinnamon cart intensifies, not dissipates, with time.

Do these happenings seem strange?

In this world, the passage of time brings increasing order. Order is the law of nature, the universal trend, the cosmic direction. If time is an arrow, that arrow points toward order. The future is pattern, organization, union, intensification; the past, randomness, confusion, disintegration, dissipation.

Philosophers have argued that without a trend toward order, time would lack meaning. The future would be indistinguishable from the past. Sequences of events would be just so many random scenes from a thousand novels. History would be indistinct, like the mist slowly gathered by treetops in evening.

In such a world, people with untidy houses lie in their beds and wait for the forces of nature to jostle the dust from their windowsills and straighten the shoes in their closets. People with untidy affairs may picnic while their calendars become organized, their appointments arranged, their accounts balanced. Lipsticks and brushes and letters may be tossed into purses with the satisfaction that they will sort themselves out automatically. Gardens need never be pruned, weeds never uprooted. Desks become neat by the end of the day. Clothes on the floor in the evening lie on chairs in the morning. Missing socks reappear.

If one visits a city in spring, one sees another wondrous sight. For in springtime the populace become sick of the order in their lives. In spring, people furiously lay waste to their houses. They sweep in dirt, smash chairs, break windows. On Aarbergergasse, or any residential avenue in spring, one hears the sounds of broken glass, shouting, howling, laughter. In spring, people meet at unarranged times, burn their appointment books, throw away their watches, drink through the night. This hysterical abandon continues until summer, when people regain their senses and return to order.

Why Does Disorder Increase in the Same Direction of Time as That in Which the Universe Expands?*

Roald Hoffman

It has something to do
with looking down the blouse
of the girl painting the boat, tracing
in a second the curve, wanting
to slip a hand between cotton
and her warm skin.

Or seeing a glint of sun
off the window opening across
the bay, calculating the speed
with which the reflection
skims across water.

The girl runs her hand
through her hair, the immemorial
action, this time arrested
as she spots the hummingbird
taking its hovering time
to sample each larkspur blossom.

*See S. Hawking, *New Scientist,* July 9, 1987, p. 46.

Or the oil storage tanks
across the water, seeing
them ignite,
silently, the shrapnel
already on its way here.

The Golden Man

Philip K. Dick

"Is it always hot like this?" the salesman demanded. He addressed everybody at the lunch counter and in the shabby booths against the wall. A middle-aged fat man with a good-natured smile, rumpled gray suit, sweat stained white shirt, a drooping bowtie, and a Panama hat.

"Only in the summer," the waitress answered.

None of the others stirred. The teen-age boy and girl in one of the booths, eyes fixed intently on each other. Two workmen, sleeves rolled up, arms dark and hairy, eating bean soup and rolls. A lean, weathered farmer. An elderly businessman in a blue-serge suit, vest and pocket watch. A dark rat-faced cab driver drinking coffee. A tired woman who had come in to get off her feet and put down her bundles.

The salesman got out a package of cigarettes. He glanced curiously around the dingy cafe, lit up, leaned his arms on the counter, and said to the man next to him: "What's the name of this town?"

The man grunted. "Walnut Creek."

The salesman sipped at his coke for a while, cigarette held loosely between plump white fingers. Presently he reached in his coat and brought out a leather wallet. For a long time he leafed thoughtfully through cards and papers, bits of notes, ticket stubs, endless odds and ends, soiled fragments—and finally a photograph.

He grinned at the photograph, and then began to chuckle,

a low moist rasp. "Look at this," he said to the man beside him.

The man went on reading his newspaper.

"Hey, look at this." The salesman nudged him with his elbow and pushed the photograph at him. "How's that strike you?"

Annoyed, the man glanced briefly at the photograph. It showed a nude woman, from the waist up. Perhaps thirty-five years old. Face turned away. Body white and flabby. With eight breasts.

"Ever seen anything like that?" the salesman chuckled, his little red eyes dancing. His face broke into lewd smiles and again he nudged the man.

"I've seen that before." Disgusted, the man resumed reading his newspaper.

The salesman noticed the lean old farmer was looking at the picture. He passed it genially over to him. "How's that strike you, pop? Pretty good stuff, eh?"

The farmer examined the picture solemnly. He turned it over, studied the creased back, took a second look at the front, then tossed it to the salesman. It slid from the counter, turned over a couple of times, and fell to the floor face up.

The salesman picked it up and brushed it off. Carefully, almost tenderly, he restored it to his wallet. The waitress's eyes flickered as she caught a glimpse of it.

"Damn nice," the salesman observed, with a wink. "Wouldn't you say so?"

The waitress shrugged indifferently. "I don't know. I saw a lot of them around Denver. A whole colony."

"That's where this was taken. Denver DCA Camp."

"Any still alive?" the farmer asked.

The salesman laughed harshly. "You kidding?" He made a short, sharp swipe with his hand. "Not any more."

They were all listening. Even the high school kids in the booth had stopped holding hands and were sitting up straight, eyes wide with fascination.

"Saw a funny kind down near San Diego," the farmer said.

"Last year, some time. Had wings like a bat. Skin, not feathers. Skin and bone wings."

The rat-eyed taxi driver chimed in. "That's nothing. There was a two-headed one in Detroit. I saw it on exhibit."

"Was it alive?" the waitress asked.

"No. They'd already euthed it."

"In sociology," the high school boy spoke up, "we saw tapes of a whole lot of them. The winged kind from down south, the big-headed one they found in Germany, an awful-looking one with sort of cones, like an insect. And—"

"The worst of all," the elderly businessman stated, "are those English ones. That hid out in the coal mines. The ones they didn't find until last year." He shook his head. "Forty years, down there in the mines, breeding and developing. Almost a hundred of them. Survivors from a group that went underground during the War."

"They just found a new kind in Sweden," the waitress said. "I was reading about it. Controls minds at a distance, they said. Only a couple of them. The DCA got there plenty fast."

"That's a variation of the New Zealand type," one of the workmen said. "It read minds."

"Reading and controlling are two different things," the businessman said. "When I hear something like that I'm plenty glad there's the DCA."

"There was a type they found right after the War," the farmer said. "In Siberia. Had the ability to control objects. Psychokinetic ability. The Soviet DCA got it right away. Nobody remembers that any more."

"I remember that," the businessman said. "I was just a kid, then. I remember because that was the first deeve I ever heard of. My father called me into the living room and told me and my brothers and sisters. We were still building the house. That was in the days when the DCA inspected everyone and stamped their arms." He held up his thin, gnarled wrist. "I was stamped there, sixty years ago."

"Now they just have the birth inspection," the waitress said. She shivered. "There was one in San Francisco this month. First in over a year. They thought it was over, around here."

"It's been dwindling," the taxi driver said. "Frisco wasn't too bad hit. Not like some. Not like Detroit."

"They still get ten or fifteen a year in Detroit," the high school boy said. "All around there. Lots of pools still left. People go into them, in spite of the robot signs."

"What kind was this one?" the salesman asked. "The one they found in San Francisco."

The waitress gestured. "Common type. The kind with no toes. Bent-over. Big eyes."

"The nocturnal type," the salesman said.

"The mother had hid it. They say it was three years old. She got the doctor to forge the DCA chit. Old friend of the family."

The salesman had finished his coke. He sat playing idly with his cigarettes, listening to the hum of talk he had set into motion. The high school boy was leaning excitedly toward the girl across from him, impressing her with his fund of knowledge. The lean farmer and the businessman were huddled together, remembering the old days, the last years of the War, before the first Ten-Year Reconstruction Plan. The taxi driver and the two workmen were swapping yarns about their own experiences.

The salesman caught the waitress's attention. "I guess," he said thoughtfully, "that one in Frisco caused quite a stir. Something like that happening so close."

"Yeah," the waitress murmured.

"This side of the Bay wasn't really hit," the salesman continued. "You never get any of them over here."

"No." The waitress moved abruptly. "None in this area. Ever." She scooped up dirty dishes from the counter and headed toward the back.

"Never?" the salesman asked, surprised. "You've never had any deeves on this side of the Bay?"

"No. None." She disappeared into the back, where the fry cook stood by his burners, white apron and tattooed wrists. Her voice was a little too loud, a little too harsh and strained. It made the farmer pause suddenly and glance up.

Silence dropped like a curtain. All sound cut off instantly. They were all gazing down at their food, suddenly tense and ominous.

"None around here," the taxi driver said, loudly and clearly, to no one in particular. "None ever."

"Sure," the salesman agreed genially. "I was only—"

"Make sure you get that straight," one of the workmen said.

The salesman blinked. "Sure, buddy. Sure." He fumbled nervously in his pocket. A quarter and a dime jangled to the floor and he hurriedly scooped them up. "No offense."

For a moment there was silence. Then the high school boy spoke up, aware for the first time that nobody was saying anything. "I heard something," he began eagerly, voice full of importance. "Somebody said they saw something up by the Johnson farm that looked like it was one of those—"

"Shut up," the businessman said, without turning his head.

Scarlet-faced, the boy sagged in his seat. His voice wavered and broke off. He peered hastily down at his hands and swallowed unhappily.

The salesman paid the waitress for his coke. "What's the quickest road to Frisco?" he began. But the waitress had already turned her back.

The people at the counter were immersed in their food. None of them looked up. They ate in frozen silence. Hostile, unfriendly faces, intent on their food.

The salesman picked up his bulging briefcase, pushed open the screen door, and stepped out into the blazing sunlight. He moved toward his battered 1978 Buick, parked a few meters up. A blue-shirted traffic cop was standing in the shade of an awning, talking languidly to a young woman in a yellow silk dress that clung moistly to her slim body.

The salesman paused a moment before he got into his car. He waved his hand and hailed the policeman. "Say, you know this town pretty good?"

The policeman eyed the salesman's rumpled gray suit, bowtie, his sweat-stained shirt. The out-of-state license. "What do you want?"

"I'm looking for the Johnson farm," the salesman said. "Here to see him about some litigation." He moved toward the policeman, a small white card between his fingers. "I'm his attorney—from the New York Guild. Can you tell me how to get out there? I haven't been through here in a couple of years."

Nat Johnson gazed up at the noonday sun and saw that it was good. He sat sprawled out on the bottom step of the porch, a pipe between his yellowed teeth, a lithe, wiry man in red-checkered shirt and canvas jeans, powerful hands, iron-gray hair that was still thick despite sixty-five years of active life.

He was watching the children play. Jean rushed laughing in front of him, bosom heaving under her sweatshirt, black hair streaming behind her. She was sixteen, bright-eyed, legs strong and straight, slim young body bent slightly forward with the weight of the two horseshoes. After her scampered Dave, fourteen, white teeth and black hair, a handsome boy, a son to be proud of. Dave caught up with his sister, passed her, and reached the far peg. He stood waiting, legs apart, hands on his hips, his two horseshoes gripped easily. Gasping, Jean hurried toward him.

"Go ahead!" Dave shouted. "You shoot first. I'm waiting for you."

"So you can knock them away?"

"So I can knock them closer."

Jean tossed down one horseshoe and gripped the other with both hands, eyes on the distant peg. Her lithe body bent, one leg slid back, her spine arched. She took careful aim, closed one eye, and then expertly tossed the shoe. With a clang the shoe struck the distant peg, circled briefly around it, then bounced off again and rolled to one side. A cloud of dust rolled up.

"Not bad," Nat Johnson admitted, from his step. "Too hard, though. Take it easy." His chest swelled with pride as the girl's glistening body took aim and again threw. Two powerful, handsome children, almost ripe, on the verge of adulthood. Playing together in the hot sun.

And there was Cris.

Cris stood by the porch, arms folded. He wasn't playing. He

was watching. He had stood there since Dave and Jean had begun playing, the same half-intent, half-remote expression on his finely-cut face. As if he were seeing past them, beyond the two of them. Beyond the field, the barn, the creek bed, the rows of cedars.

"Come on, Cris!" Jean called, as she and Dave moved across the field to collect their horseshoes. "Don't you want to play?"

No, Cris didn't want to play. He never played. He was off in a world of his own, a world into which none of them could come. He never joined in anything, games or chores or family activities. He was by himself always. Remote, detached, aloof. Seeing past everyone and everything—that is, until all at once something clicked and he momentarily rephased, reentered their world briefly.

Nat Johnson reached out and knocked his pipe against the step. He refilled it from his leather tobacco pouch, his eyes on his eldest son. Cris was now moving into life. Heading out onto the field. He walked slowly, arms folded calmly, as if he had, for the moment, descended from his own world into theirs. Jean didn't see him; she had turned her back and was getting ready to pitch.

"Hey," Dave said, startled. "Here's Cris."

Cris reached his sister, stopped, and held out his hand. A great dignified figure, calm and impassive. Uncertainly, Jean gave him one of the horseshoes. "You want this? You want to play?"

Cris said nothing. He bent slightly, a supple arc of his incredibly graceful body, then moved his arm in a blur of speed. The shoe sailed, struck the far peg, and dizzily spun around it. Ringer.

The corners of Dave's mouth turned down. "What a lousy darn thing."

"Cris," Jean reproved. "You don't play fair."

No, Cris didn't play fair. He had watched half an hour—then come out and thrown once. One perfect toss, one dead ringer.

"He never makes a mistake," Dave complained.

Cris stood, face blank. A golden statue in the mid-day sun. Golden hair, skin, a light down of gold fuzz on his bare arms and legs—

Abruptly he stiffened. Nat sat up, startled. "What is it?" he barked.

Cris turned in a quick circle, magnificent body alert. "Cris!" Jean demanded. "What—"

Cris shot forward. Like a released energy beam he bounded across the field, over the fence, into the barn and out the other side. His flying figure seemed to skim over the dry grass as he descended into the barren creek bed, between the cedars. A momentary flash of gold—and he was gone. Vanished. There was no sound. No motion. He had utterly melted into the scenery.

"What was it this time?" Jean asked wearily. She came over to her father and threw herself down in the shade. Sweat glowed on her smooth neck and upper lip; her sweat shirt was streaked and damp. "What did he see?"

"He was after something," Dave stated, coming up.

Nat grunted. "Maybe. There's no telling."

"I guess I better tell Mom not to set a place for him," Jean said. "He probably won't be back."

Anger and futility descended over Nat Johnson. No, he wouldn't be back. Not for dinner and probably not the next day—or the one after that. He'd be gone God only knew how long. Or where. Or why. Off by himself, alone some place. "If I thought there was any use," Nat began, "I'd send you two after him. But there's no—"

He broke off. A car was coming up the dirt road toward the farmhouse. A dusty, battered old Buick. Behind the wheel sat a plump red-faced man in a gray suit, who waved cheerfully at them as the car sputtered to a stop and the motor died into silence.

"Afternoon," the man nodded, as he climbed out the car. He tipped his hat pleasantly. He was middle-aged, genial-looking, perspiring freely as he crossed the dry ground toward the porch. "Maybe you folks can help me."

"What do you want?" Nat Johnson demanded hoarsely. He was frightened. He watched the creek bed out of the corner of his eye, praying silently. God, if only he *stayed* away. Jean was breathing quickly, sharp little gasps. She was terrified. Dave's face was expressionless, but all color had drained from it. "Who are you?" Nat demanded.

"Name's Baines. George Baines." The man held out his hand but Johnson ignored it. "Maybe you've heard of me. I own the Pacifica Development Corporation. We built all those little bomb-proof houses just outside town. Those little round ones you see as you come up the main highway from Lafayette."

"What do you want?" Johnson held his hands steady with an effort. He'd never heard of the man, although he'd noticed the housing tract. It couldn't be missed—a great ant-heap of ugly pill-boxes straddling the highway. Baines looked like the kind of man who'd own them. But what did he want here?

"I've bought some land up this way," Baines was explaining. He rattled a sheaf of crisp papers. "This is the deed, but I'll be damned if I can find it." He grinned good-naturedly. "I know it's around this way, someplace, this side of the State road. According to the clerk at the County Recorder's Office, a mile or so this side of that hill over there. But I'm no damn good at reading maps."

"It isn't around here," Dave broke in. "There's only farms around here. Nothing for sale."

"This is a farm, son," Baines said genially. "I bought it for myself and my missus. So we could settle down." He wrinkled his pug nose. "Don't get the wrong idea—I'm not putting up any tracts around here. This is strictly for myself. An old farmhouse, twenty acres, a pump and a few oak trees—"

"Let me see the deed." Johnson grabbed the sheaf of papers, and while Baines blinked in astonishment, he leafed rapidly through them. His face hardened and he handed them back. "What are you up to? This deed is for a parcel fifty miles from here."

"Fifty miles!" Baines was dumbfounded. "No kidding? But the clerk told me—"

Johnson was on his feet. He towered over the fat man. He was in top-notch physical shape—and he was plenty damn suspicious. "Clerk, hell. You get back into your car and drive out of here. I don't know what you're after, or what you're here for, but I want you off my land."

In Johnson's massive fist something sparkled. A metal tube that gleamed ominously in the mid-day sunlight. Baines saw it—and gulped. "No offense, mister." He backed nervously away. "You folks sure are touchy. Take it easy, will you?"

Johnson said nothing. He gripped the lash-tube tighter and waited for the fat man to leave.

But Baines lingered. "Look, buddy. I've been driving around this furnace five hours, looking for my damn place. Any objection to my using your—facilities?"

Johnson eyed him with suspicion. Gradually the suspicion turned to disgust. He shrugged. "Dave, show him where the bathroom is."

"Thanks." Baines grinned thankfully. "And if it wouldn't be too much trouble, maybe a glass of water. I'd be glad to pay you for it." He chuckled knowingly. "Never let the city people get away with anything, eh?"

"Christ." Johnson turned away in revulsion as the fat man lumbered after his son, into the house.

"Dad," Jean whispered. As soon as Baines was inside she hurried up onto the porch, eyes wide with fear. "Dad, do you think he—"

Johnson put his arm around her. "Just hold on tight. He'll be gone, soon."

The girl's dark eyes flashed with mute terror. "Every time the man from the water company, or the tax collector, some tramp, children, *anybody* come around, I get a terrible stab of pain—here." She clutched at her heart, hand against her breasts. "It's been that way thirteen years. How much longer can we keep it going? *How long?*"

The man named Baines emerged gratefully from the bathroom. Dave Johnson stood silently by the door, body rigid, youthful face stony.

"Thanks, son," Baines sighed. "Now where can I get a glass of cold water?" He smacked his thick lips in anticipation. "After you've been driving around the sticks looking for a dump some red-hot real estate agent stuck you with—"

Dave headed into the kitchen. "Mom, this man wants a drink of water. Dad said he could have it."

Dave had turned his back. Baines caught a brief glimpse of the mother, gray-haired, small, moving toward the sink with a glass, face withered and drawn, without expression.

Then Baines hurried from the room down a hall. He passed through a bedroom, pulled a door open, found himself facing a closet. He turned and raced back, through the living room, into a dining room, then another bedroom. In a brief instant he had gone through the whole house.

He peered out a window. The back yard. Remains of a rusting truck. Entrance of an underground bomb shelter. Tin cans. Chickens scratching around. A dog, asleep under a shed. A couple of old auto tires.

He found a door leading out. Soundlessly, he tore the door open and stepped outside. No one was in sight. There was the barn, a leaning, ancient wood structure. Cedar trees beyond, a creek of some kind. What had once been an outhouse.

Baines moved cautiously around the side of the house. He had perhaps thirty seconds. He had left the door of the bathroom closed; the boy would think he had gone back in there. Baines looked into the house through a window. A large closet, filled with old clothing, boxes and bundles of magazines.

He turned and started back. He reached the corner of the house and started around it.

Nat Johnson's gaunt shape loomed up and blocked his way. "All right, Baines. You asked for it."

A pink flash blossomed. It shut out the sunlight in a single blinding burst. Baines leaped back and clawed at his coat pocket. The edge of the flash caught him and he half-fell, stunned by the force. His suit-shield sucked in the energy and discharged it, but the power rattled his teeth and for a moment he jerked like a puppet on a string. Darkness ebbed around him. He could feel the mesh of the shield glow white, as it absorbed the energy and fought to control it.

His own tube came out—and Johnson had no shield. "You're under arrest," Baines muttered grimly. "Put down your tube and your hands up. And call your family." He made a motion with the tube. "Come on, Johnson. Make it snappy."

The lash-tube wavered and then slipped from Johnson's fingers. "You're still alive." Dawning horror crept across his face. "Then you must be—"

Dave and Jean appeared. *"Dad!"*

"Come over here," Baines ordered. "Where's your mother?"

Dave jerked his head numbly. "Inside."

"Get her and bring her here."

"You're DCA," Nat Johnson whispered.

Baines didn't answer. He was doing something with his neck, pulling at the flabby flesh. The wiring of a contact mike glittered as he slipped it from a fold between two chins and into his pocket. From the dirt road came the sound of motors, sleek purrs that rapidly grew louder. Two teardrops of black metal came gliding up and parked beside the house. Men swarmed out, in the dark gray-green of the Government Civil Police. In the sky swarms of black dots were descending, clouds of ugly flies that darkened the sun as they spilled out men and equipment. The men drifted slowly down.

"He's not here," Baines said, as the first man reached him. "He got away. Inform Wisdom back at the lab."

"We've got this section blocked off."

Baines turned to Nat Johnson, who stood in dazed silence, uncomprehending, his son and daughter beside him. "How did he know we were coming?" Baines demanded.

"I don't know," Johnson muttered. "He just—knew."

"A telepath?"

"I don't know."

Baines shrugged. "We'll know, soon. A clamp is out, all around here. He can't get past, no matter what the hell he can do. Unless he can dematerialize himself."

"What'll you do with him when you—if you catch him?" Jean asked huskily.

"Study him."

"And then kill him?"

"That depends on the lab evaluation. If you could give me more to work on, I could predict better."

"We can't tell you anything. We don't know anything more." The girl's voice rose with desperation. "He doesn't talk."

Baines jumped. *"What?"*

"He doesn't talk. He never talked to us. Ever."

"How old is he?"

"Eighteen."

"No communication." Baines was sweating. "In eighteen years there hasn't been any semantic bridge between you? Does he have *any* contact? Signs? Codes?"

"He—ignores us. He eats here, stays with us. Sometimes he plays when we play. Or sits with us. He's gone days on end. We've never been able to find out what he's doing—or where. He sleeps in the barn—by himself."

"Is he really gold-colored?"

"Yes. Skin, eyes, hair, nails. Everything."

"And he's large? Well-formed?"

It was a moment before the girl answered. A strange emotion stirred her drawn features, a momentary glow. "He's incredibly beautiful. A god come down to earth." Her lips twisted. "You won't find him. He can do things. Things you have no comprehension of. Powers so far beyond your limited—"

"You don't think we'll get him?" Baines frowned. "More teams are landing all the time. You've never seen an Agency clamp in operation. We've had sixty years to work out all the bugs. If he gets away it'll be the first time—"

Baines broke off abruptly. Three men were quickly approaching the porch. Two green-clad Civil Police. And a third man between them. A man who moved silently, lithely, a faintly luminous shape that towered above them.

"Cris!" Jean screamed.

"We got him," one of the police said.

Baines fingered his lash-tube uneasily. "Where? How?"

"He gave himself up," the policeman answered, voice full of

awe. "He came to us voluntarily. Look at him. He's like a metal statue. Like some sort of—god."

The golden figure halted for a moment beside Jean. Then it turned slowly, calmly, to face Baines.

"Cris!" Jean shrieked. *"Why did you come back?"*

The same thought was eating at Baines, too. He shoved it aside—for the time being. "Is the jet out front?" he demanded quickly.

"Ready to go," one of the CP answered.

"Fine." Baines strode past them, down the steps and onto the dirt field. "Let's go. I want him taken directly to the lab." For a moment he studied the massive figure who stood calmly between the two Civil Policemen. Beside him, they seemed to have shrunk, become ungainly and repellent. Like dwarves . . . What had Jean said? *A god come to earth.* Baines broke angrily away. "Come on," he muttered brusquely. "This one may be tough; we've never run up against one like it before. We don't know what the hell it can do."

The chamber was empty, except for the seated figure. Four bare walls, floor and ceiling. A steady glare of white light relentlessly etched every corner of the chamber. Near the top of the far wall ran a narrow slot, the view windows through which the interior of the chamber was scanned.

The seated figure was quiet. He hadn't moved since the chamber locks had slid into place, since the heavy bolts had fallen from outside and the rows of bright-faced technicians had taken their places at the view windows. He gazed down at the floor, bent forward, hands clasped together, face calm, almost expressionless. In four hours he hadn't moved a muscle.

"Well?" Baines said. "What have you learned?"

Wisdom grunted sourly. "Not much. If we don't have him doped out in forty-eight hours we'll go ahead with the euth. We can't take any chances."

"You're thinking about the Tunis type," Baines said. He was, too. They had found ten of them, living in the ruins of the abandoned North African town. Their survival method was simple.

They killed and absorbed other life forms, then imitated them and took their places. *Chameleons*, they were called. It had cost sixty lives, before the last one was destroyed. Sixty top-level experts, highly trained DCA men.

"Any clues?" Baines asked.

"He's different as hell. This is going to be tough." Wisdom thumbed a pile of tape-spools. "This is the complete report, all the material we got from Johnson and his family. We pumped them with the psych-wash, then let them go home. Eighteen years—and no semantic bridge. Yet, he looks fully developed. Mature at thirteen—a shorter, faster life-cycle than ours. But why the mane? All the gold fuzz? Like a Roman monument that's been gilded."

"Has the report come in from the analysis room? You had a wave-shot taken, of course."

"His brain pattern has been fully scanned. But it takes time for them to plot it out We're all running around like lunatics while he just sits there!" Wisdom poked a stubby finger at the window "We caught him easily enough. He can't have *much*, can he? But I'd like to know what it is. Before we euth him."

"Maybe we should keep him alive until we know."

"Euth in forty-eight hours," Wisdom repeated stubbornly. "Whether we know or not. I don't like him. He gives me the creeps."

Wisdom stood chewing nervously on his cigar, a red-haired, beefy-faced man, thick and heavy-set, with a barrel chest and cold, shrewd eyes deep-set in his hard face. Ed Wisdom was Director of DCA's North American Branch. But right now he was worried. His tiny eyes darted back and forth, alarmed flickers of gray in his brutal, massive face.

"You think," Baines said slowly, "this is *it?*"

"I always think so," Wisdom snapped. "I have to think so."

"I mean—"

"I know what you mean." Wisdom paced back and forth, among the study tables, technicians at their benches, equipment and humming computers. Buzzing tape-slots and research hookups. "This thing lived eighteen years with his family and *they* don't understand it. *They* don't know what it has. They know what it does, but not how."

"What does it do?"

"It knows things."

"What kind of things?"

Wisdom grabbed his lash-tube from his belt and tossed it on a table. "Here."

"What?"

"Here." Wisdom signalled, and a view window was slid back an inch. "Shoot him."

Baines blinked. "You said forty-eight hours."

With a curse, Wisdom snatched up the tube, aimed it through the window directly at the seated figure's back, and squeezed the trigger.

A blinding flash of pink. A cloud of energy blossomed in the center of the chamber. It sparkled, then died into dark ash.

"Good God!" Baines gasped. "You—"

He broke off. The figure was no longer sitting. As Wisdom fired, it had moved in a blur of speed, away from the blast, to the corner of the chamber. Now it was slowly coming back, face blank, still absorbed in thought.

"Fifth time," Wisdom said, as he put his tube away. "Last time Jamison and I fired together. Missed. He knew exactly when the bolts would hit. And where."

Baines and Wisdom looked at each other. Both of them were thinking the same thing. "But even reading minds wouldn't tell him where they were going to hit," Baines said. "When, maybe. But not where. Could you have called your own shots?"

"Not mine," Wisdom answered flatly. "I fired fast, damn near at random." He frowned. "*Random*. We'll have to make a test of this." He waved a group of technicians over. "Get a construction team up here. On the double." He grabbed paper and pen and began sketching.

While construction was going on, Baines met his fiancée in the lobby outside the lab, the great central lounge of the DCA Building.

"How's it coming?" she asked. Anita Ferris was tall and blonde, blue eyes and a mature, carefully cultivated figure. An attractive, competent-looking woman in her late twenties. She wore a metal foil dress and cape—with a red and black stripe on the sleeve, the emblem of the A-Class. Anita was Director of the Semantics Agency, a top-level Government Coordinator. "Anything of interest, this time?"

"Plenty." Baines guided her from the lobby, into the dim recess of the bar. Music played softly in the background, a shifting variety of patterns formed mathematically. Dim shapes moved expertly through the gloom, from table to table. Silent, efficient robot waiters.

As Anita sipped her Tom Collins, Baines outlined what they had found.

"What are the chances," Anita asked slowly, "that he's built up some kind of deflection-cone? There was one kind that warped their environment by direct mental effort. No tools. Direct mind to matter."

"Psychokinetics?" Baines drummed restlessly on the table top. "I doubt it. The thing has ability to predict, not to control. He can't stop the beams, but he can sure as hell get out of the way."

"Does he jump between the molecules?"

Baines wasn't amused. "This is serious. We've handled these things sixty years—longer than you and I have been around added together. Eighty-seven types of deviants have shown up, real mutants that could reproduce themselves, not mere freaks. This is the eighty-eighth. We've been able to handle each of them in turn. But this—"

"Why are you so worried about this one?"

"First, it's eighteen years old. That in itself is incredible. Its family managed to hide it that long."

"Those women around Denver were older than that. Those ones with—"

"They were in a Government camp. Somebody high up was toying with the idea of allowing them to breed. Some sort of industrial use. We withheld euth for years. But Cris Johnson stayed alive *outside our control.* Those things at Denver were under constant scrutiny."

"Maybe he's harmless. You always assume a deeve is a menace. He might even be beneficial. Somebody thought those women might work in. Maybe this thing has something that would advance the race."

"*Which* race? Not the human race. It's the old 'the operation was a success but the patient died' routine. If we introduce a mutant to keep us going it'll be mutants, not us, who'll inherit the earth. It'll be mutants surviving for their own sake. Don't think for a moment we can put padlocks on them and expect them to serve us. If they're really superior to homo sapiens, they'll win out in even competition. To survive, we've got to cold-deck them right from the start."

"In other words, we'll know homo superior when he comes—by definition. He'll be the one we won't be able to euth."

"That's about it," Baines answered. "Assuming there *is* a homo superior. Maybe there's just homo peculiar. Homo with an improved line."

"The Neanderthal probably thought the Cro-Magnon man had merely an improved line. A little more advanced ability to conjure up symbols and shape flint. From your description, this thing is more radical than a mere improvement."

"This thing," Baines said slowly, "has an ability to predict. So far, it's been able to stay alive. It's been able to cope with situations better than you or I could. How long do you think we'd stay alive in that chamber, with energy beams blazing down at us? In a sense it's got the ultimate survival ability. If it can always be accurate—"

A wall-speaker sounded. "Baines, you're wanted in the lab. Get the hell out of the bar and upramp."

Baines pushed back his chair and got to his feet. "Come along. You may be interested in seeing what Wisdom has got dreamed up."

A tight group of top-level DCA officials stood around in a circle, middle-aged, gray-haired, listening to a skinny youth in a white shirt and rolled-up sleeves explaining an elaborate cube of metal and plastic that filled the center of the view-platform. From it jutted an ugly array of tube snouts, gleaming muzzles that disappeared into an intricate maze of wiring.

"This," the youth was saying briskly, "is the first real test. It fires at random—as nearly random as we can make it, at least. Weighted balls are thrown up in an air stream, then dropped free to fall back and cut relays. They can fall in almost any pattern. The thing fires according to their pattern. Each drop produces a new configuration of timing and position. Ten tubes, in all. Each will be in constant motion."

"And *nobody* knows how they'll fire?" Anita asked.

"Nobody." Wisdom rubbed his thick hands together. "Mind reading won't help him, not with this thing."

Anita moved over to the view windows, as the cube was rolled into place. She gasped. "Is that him?"

"What's wrong?" Baines asked.

Anita's cheeks were flushed. "Why, I expected a—a *thing*. My God, he's beautiful! Like a golden statue. Like a deity!"

Baines laughed. "He's eighteen years old, Anita. Too young for you."

The woman was still peering through the view window. "Look at him. Eighteen? I don't believe it."

Cris Johnson sat in the center of the chamber, on the floor. A posture of contemplation, head bowed, arms folded, legs tucked under him. In the stark glare of the overhead lights his powerful body glowed and rippled, a shimmering figure of downy gold.

"Pretty, isn't he?" Wisdom muttered. "All right. Start it going."

"You're going to *kill* him?" Anita demanded.

"We're going to try."

"But he's—" She broke off uncertainly. "He's not a monster. He's not like those others, those hideous things with two heads, or those insects. Or those awful things from Tunis."

"What is he, then?" Baines asked.

"I don't know. But you can't just *kill* him. It's terrible!"

The cube clicked into life. The muzzles jerked, silently altered position. Three retracted, disappeared into the body of the cube. Others came out. Quickly, efficiently, they moved into position—and abruptly, without warning, opened fire.

A staggering burst of energy fanned out, a complex pattern that altered each moment, different angles, different velocities, a bewildering blur that cracked from the windows down into the chamber.

The golden figure moved. He dodged back and forth, expertly avoiding the bursts of energy that seared around him on all sides. Rolling clouds of ash obscured him; he was lost in a mist of crackling fire and ash.

"Stop it!" Anita shouted. "For God's sake, you'll destroy him!"

The chamber was an inferno of energy. The figure had completely disappeared. Wisdom waited a moment, then nodded to the technicians operating the cube. They touched guide buttons and the muzzles slowed and died. Some sank back into the cube. All became silent. The works of the cube ceased humming.

Cris Johnson was still alive. He emerged from the settling clouds of ash, blackened and singed. But unhurt. He had avoided each beam. He had weaved between them and among them as they came, a dancer leaping over glittering sword-points of pink fire. He had survived.

"No," Wisdom murmured, shaken and grim. "Not a telepath. Those were at random. No prearranged pattern."

The three of them looked at each other, dazed and frightened. Anita was trembling. Her face was pale and her blue eyes were wide. "What, then?" she whispered. "What is it? What does he have?"

"He's a good guesser," Wisdom suggested.

"He's not guessing," Baines answered. "Don't kid yourself. That's the whole point."

"No, he's not guessing." Wisdom nodded slowly. "He *knew*. He predicted each strike. I wonder . . . *Can* he err? *Can* he make a mistake?"

"We caught him," Baines pointed out.

"You said he came back voluntarily." There was a strange look on Wisdom's face. "Did he come back *after* the clamp was up?"

Baines jumped. "Yes, after."

"He couldn't have got through the clamp. So he came back." Wisdom grinned wryly. "The clamp must actually have been perfect. It was supposed to be."

"If there had been a single hole," Baines murmured, "he would have known it—gone through."

Wisdom ordered a group of armed guards over. "Get him out of there. To the euth stage."

Anita shrieked. "Wisdom, you can't—"

"He's too far ahead of us. We can't compete with him." Wisdom's eyes were bleak. "We can only guess what's going to happen. *He knows.* For him, it's a sure thing. I don't think it'll help him at euth, though. The whole stage is flooded simultaneously. Instantaneous gas, released throughout." He signalled impatiently to the guards. "Get going. Take him down right away. Don't waste any time."

"Can we?" Baines murmured thoughtfully.

The guards took up positions by one of the chamber locks. Cautiously, the tower control slid the lock back. The first two guards stepped cautiously in, lash-tubes ready.

Cris stood in the center of the chamber. His back was to them as they crept toward him. For a moment he was silent, utterly unmoving. The guards fanned out, as more of them entered the chamber. Then—

Anita screamed. Wisdom cursed. The golden figure spun and leaped forward, in a flashing blur of speed. Past the triple line of guards, through the lock and into the corridor.

"Get him!" Baines shouted.

Guards milled everywhere. Flashes of energy lit up the corridor, as the figure raced among them up the ramp.

"No use," Wisdom said calmly. "We can't hit him." He touched a button, then another. "But maybe this will help."

"What—" Baines began. But the leaping figure shot abruptly at him, straight at him, and he dropped to one side. The figure flashed past. It ran effortlessly, face without expression, dodging and jumping as the energy beams seared around it.

For an instant the golden face loomed up before Baines. It passed and disappeared down a side corridor. Guards rushed after it, kneeling and firing, shouting orders excitedly. In the bowels of the building, heavy guns were rumbling up. Locks slid into place as escape corridors were systematically sealed off.

"Good God," Baines gasped, as he got to his feet. "Can't he do anything but run?"

"I gave orders," Wisdom said, "to have the building isolated. There's no way out. Nobody comes and nobody goes. He's loose here in the building—but he won't get out."

"If there's one exit overlooked, he'll know it," Anita pointed out shakily.

"We won't overlook any exit. We got him once; we'll get him again."

A messenger robot had come in. Now it presented its message respectfully to Wisdom. "From analysis, sir."

Wisdom tore the tape open. "Now we'll know how it thinks." His hands were shaking. "Maybe we can figure out its blind spot. It may be able to out-think us, but that doesn't mean it's invulnerable. It only predicts the future—it can't change it. If there's only death ahead, its ability won't . . ."

Wisdom's voice faded into silence. After a moment he passed the tape to Baines.

"I'll be down in the bar," Wisdom said. "Getting a good stiff drink." His face had turned lead-gray. "All I can say is *I hope to hell this isn't the race to come.*"

"What's the analysis?" Anita demanded impatiently, peering over Baines' shoulder. "How does it think?"

"It doesn't," Baines said, as he handed the tape back to his boss. "It doesn't think at all. Virtually no frontal lobe. It's not a human being—it doesn't use symbols. It's nothing but an animal."

"An animal," Wisdom said. "With a single highly-developed faculty. Not a superior man. Not a man at all."

Up and down the corridors of the DCA Building, guards and equipment clanged. Loads of Civil Police were pouring into the building and taking up positions beside the guards. One by one, the corridors and rooms were being inspected and sealed off. Sooner or later the golden figure of Cris Johnson would be located and cornered.

"We were always afraid a mutant with superior intellectual powers would come along," Baines said reflectively. "A deeve who would be to us what we are to the great apes. Something with a bulging cranium, telepathic ability, a perfect semantic system, ultimate powers of symbolization and calculation. A development along our own path. A better human being."

"He acts by reflex," Anita said wonderingly. She had the analysis and was sitting at one of the desks studying it intently. "Reflex—like a lion. A golden lion." She pushed the tape aside, a strange expression on her face. "The lion god."

"Beast," Wisdom corrected tartly. "Blond beast, you mean"

"He runs fast," Baines said, "and that's all. No tools. He doesn't build anything or utilize anything outside himself. He just stands and waits for the right opportunity and then he runs like hell."

"This is worse than anything we've anticipated," Wisdom said. His beefy face was lead-gray. He sagged like an old man, his blunt hands trembling and uncertain. "To be replaced by an animal. Something that runs and hides. Something without a language!" He spat savagely. "That's why they weren't able to communicate with it. We wondered what kind of semantic system it had. It hasn't got any! No more ability to talk and think than a—dog."

"That means intelligence has failed," Baines went on huskily. "We're the last of our line—like the dinosaur. We've carried intelligence as far as it'll go. Too far, maybe. We've already got to the point where we know so much—think so much—we can't act."

"Men of thought," Anita said. "Not men of action. It's begun to have a paralyzing effect. But this thing—"

"This thing's faculty works better than ours ever did. We can recall past experiences, keep them in mind, learn from them. At best, we can make shrewd guesses about the future, from our memory of what's happened in the past. But we can't be certain. We have to speak of probabilities. Grays. Not blacks and whites. We're only guessing."

"Cris Johnson isn't guessing," Anita added.

"He can look ahead. See what's coming. He can—prethink. Let's call it that. He can see into the future. Probably he doesn't perceive it as the future."

"No," Anita said thoughtfully. "It would seem like the present. He has a broader present. But his present lies ahead, not back. Our present is related to the past. Only the past is certain, to us. To him, the future is certain. And he probably doesn't remember the past, any more than any animal remembers what happened."

"As he develops," Baines said, "as his race evolves, it'll probably expand its ability to prethink. Instead of ten minutes, thirty minutes. Then an hour. A day. A year. Eventually they'll be able to keep ahead a whole lifetime. Each one of them will live in a solid, unchanging world. There'll be no variables, no uncertainty. No motion! They won't have anything to fear. Their world will be perfectly static, a solid block of matter."

"And when death comes," Anita said, "they'll accept it. There won't be any struggle; to them, it'll already have happened."

"Already have happened," Baines repeated. "To Cris, our shots had already been fired." He laughed harshly. "Superior survival doesn't mean superior man. If there were another world-wide flood, only fish would survive. If there were another ice age, maybe nothing but polar bears would be left. When we opened the lock, he had already seen the men, seen exactly where they were standing and what they'd do. A neat faculty—but not a development of mind. A pure physical *sense.*"

"But if every exit is covered," Wisdom repeated, "he'll see he can't get out. He gave himself up before—he'll give himself up again." He shook his head. "An animal. Without language. Without tools."

"With his new sense," Baines said, "he doesn't need anything else." He examined his watch. "It's after two. Is the building completely sealed off?"

"You can't leave," Wisdom stated. "You'll have to stay here all night—or until we catch the bastard."

"I meant her." Baines indicated Anita. "She's supposed to be back at Semantics by seven in the morning."

Wisdom shrugged. "I have no control over her. If she wants, she can check out."

"I'll stay," Anita decided. "I want to be here when he—when he's destroyed. I'll sleep here." She hesitated. "Wisdom, isn't there some other way? If he's just an animal couldn't we—"

"A zoo?" Wisdom's voice rose in a frenzy of hysteria. "Keep it penned up in the zoo? Christ no! It's got to be killed!"

For a long time the great gleaming shape crouched in the darkness. He was in a store room. Boxes and cartons stretched out on all sides, heaped up in orderly rows, all neatly counted and marked. Silent and deserted.

But in a few moments people burst in and searched the room. He could see this. He saw them in all parts of the room, clear and distinct, men with lash-tubes, grim-faced, stalking with murder in their eyes.

The sight was one of many. One of a multitude of clearly-etched scenes lying tangent to his own. And to each was attached a further multitude of interlocking scenes, that finally grew hazier and dwindled away. A progressive vagueness, each syndrome less distinct.

But the immediate one, the scene that lay closest to him, was clearly visible. He could easily make out the sight of the armed men. Therefore it was necessary to be out of the room before they appeared.

The golden figure got calmly to its feet and moved to the door. The corridor was empty; he could see himself already outside, in the vacant, drumming hall of metal and recessed lights. He pushed the door boldly open and stepped out.

A lift blinked across the hall. He walked to the lift and entered it. In five minutes a group of guards would come running along and leap into the lift. By that time he would have left it and sent it back down. Now he pressed a button and rose to the next floor.

He stepped out into a deserted passage. No one was in sight. That didn't surprise him. He couldn't be surprised. The element didn't exist for him. The positions of things, the space relationships of all matter in the immediate future, were as certain for him as his own body. The only thing that was unknown was that which had already passed out of being. In a vague, dim fashion, he had occasionally wondered where things went after he had passed them.

He came to a small supply closet. It had just been searched. It would be a half an hour before anyone opened it again. He had that long; he could see that far ahead. And then—

And then he would be able to see another area, a region farther beyond. He was always moving, advancing into new regions he had never seen before. A constantly unfolding panorama of sights and scenes, frozen landscapes spread out ahead. All objects were fixed. Pieces on a vast chess board through which he moved, arms folded, face calm. A detached observer who saw objects that lay ahead of him as clearly as those under foot.

Right now, as he crouched in the small supply closet, he saw an unusually varied multitude of scenes for the next half hour. Much lay ahead. The half hour was divided into an incredibly complex pattern of separate configurations. He had reached a critical region; he was about to move through worlds of intricate complexity.

He concentrated on a scene ten minutes away. It showed, like a three dimensional still, a heavy gun at the end of the corridor, trained all the way to the far end. Men moved cautiously from door to door, checking each room again, as they had done repeatedly. At the end of the half hour they had reached the supply closet. A scene showed them looking inside. By that time he was gone, of course. He wasn't in that scene. He had passed on to another.

The next scene showed an exit. Guards stood in a solid line. No way out. He was in that scene. Off to one side, in a niche just inside the door. The street outside was visible, stars, lights, outlines of passing cars and people.

In the next tableau he had gone back, away from the exit.

There was no way out. In another tableau he saw himself at other exits, a legion of golden figures, duplicated again and again, as he explored regions ahead, one after another. But each exit was covered.

In one dim scene he saw himself lying charred and dead; he had tried to run through the line, out the exit.

But that scene was vague. One wavering, indistinct still out of many. The inflexible path along which he moved would not deviate in that direction. It would not turn him that way. The golden figure in that scene, the miniature doll in that room, was only distantly related to him. It was himself, but a far-away self. A self he would never meet. He forgot it and went on to examine the other tableau.

The myriad of tableaux that surrounded him were an elaborate maze, a web which he now considered bit by bit. He was looking down into a doll's house of infinite rooms, rooms without number, each with its furniture, its dolls, all rigid and unmoving. The same dolls and furniture were repeated in many. He, himself, appeared often. The two men on the platform. The woman. Again and again the same combinations turned up; the play was redone frequently, the same actors and props moved around in all possible ways.

Before it was time to leave the supply closet, Cris Johnson had examined each of the rooms tangent to the one he now occupied. He had consulted each, considered its contents thoroughly.

He pushed the door open and stepped calmly out into the hall. He knew exactly where he was going. And what he had to do. Crouched in the stuffy closet, he had quietly and expertly examined each miniature of himself, observed which clearly-etched configuration lay along his inflexible path, the one room of the doll house, the one set out of legions, toward which he was moving.

Anita slipped out of her metal foil dress, hung it over a hanger, then unfastened her shoes and kicked them under the bed. She was just starting to unclip her bra when the door opened.

She gasped. Soundlessly, calmly, the great golden shape closed the door and bolted it after him.

Anita snatched up her lash-tube from the dressing table. Her hand shook; her whole body was trembling. "What do you want?" she demanded. Her fingers tightened convulsively around the tube. "I'll kill you."

The figure regarded her silently, arms folded. It was the first time she had seen Cris Johnson closely. The great dignified face, handsome and impassive. Broad shoulders. The golden mane of hair, golden skin, pelt of radiant fuzz—

"Why?" she demanded breathlessly. Her heart was pounding wildly. "What do you want?"

She could kill him easily. But the lash-tube wavered. Cris Johnson stood without fear; he wasn't at all afraid. Why not? Didn't he understand what it was? What the small metal tube could do to him?

"Of course," she said suddenly, in a choked whisper. "You can see ahead. You know I'm going to kill you. Or you wouldn't have come here."

She flushed, terrified—and embarrassed. He knew exactly what she was going to do; he could see it as easily as she saw the walls of the room, the wall-bed with its covers folded neatly back, her clothes hanging in the closet, her purse and small things on the dressing table.

"All right." Anita backed away, then abruptly put the tube down on the dressing table. "I won't kill you. Why should I?" She fumbled in her purse and got out her cigarettes. Shakily, she lit up, her pulse racing. She was scared. And strangely fascinated. "Do you expect to stay here? It won't do any good. They've come through the dorm twice, already. They'll be back."

Could he understand her? She saw nothing on his face, only blank dignity. God, he was huge! It wasn't possible he was only eighteen, a boy, a child. He looked more like some great golden god, come down to earth.

She shook the thought off savagely. He wasn't a god. He was a beast. *The blond beast,* come to take the place of man. To drive man from the earth.

Anita snatched up the lash-tube. "Get out of here! You're an animal! A big stupid animal! You can't even understand what I'm saying—you don't even have a language. You're not human."

Cris Johnson remained silent. As if he were waiting. Waiting for what? He showed no sign of fear or impatience, even though the corridor outside rang with the sound of men searching, metal against metal, guns and energy tubes being dragged around, shouts and dim rumbles as section after section of the building was searched and sealed off.

"They'll get you," Anita said. "You'll be trapped here. They'll be searching this wing any moment." She savagely stubbed out her cigarette. "For God's sake, what do you expect *me* to do?"

Cris moved toward her. Anita shrank back. His powerful hands caught hold of her and she gasped in sudden terror. For a moment she struggled blindly, desperately.

"Let go!" She broke away and leaped back from him. His face was expressionless. Calmly, he came toward her, an impassive god advancing to take her. "Get away!" She groped for the lash-tube, trying to get up. But the tube slipped from her fingers and rolled onto the floor.

Cris bent down and picked it up. He held it out to her, in the open palm of his hand.

"Good God," Anita whispered. Shakily, she accepted the tube, gripped it hesitantly, then put it down again on the dressing table.

In the half-light of the room, the great golden figure seemed to glow and shimmer, outlined against the darkness. A god—no, not a god. An animal. A great golden beast, without a soul. She was confused. Which was he—or was he both? She shook her head, bewildered. It was late, almost four. She was exhausted and confused.

Cris took her in his arms. Gently, kindly, he lifted her face and kissed her. His powerful hands held her tight. She couldn't breathe. Darkness, mixed with the shimmering golden haze, swept around her. Around and around it spiralled, carrying her senses away. She sank down into it gratefully. The darkness covered her and dissolved her in a swelling torrent of sheer force that mounted in intensity each moment, until the roar of it beat against her and at last blotted out everything.

Anita blinked. She sat up and automatically pushed her hair into place. Cris was standing before the closet. He was reaching up, getting something down.

He turned toward her and tossed something on the bed. Her heavy metal foil traveling cape.

Anita gazed down at the cape without comprehension. "What do you want?"

Cris stood by the bed, waiting.

She picked up the cape uncertainly. Cold creepers of fear plucked at her. "You want me to get you out of here," she said softly. "Past the guards and the CP."

Cris said nothing.

"They'll kill you instantly." She got unsteadily to her feet. "You can't run past them. Good God, don't you do anything but run? There must be a better way. Maybe I can appeal to Wisdom. I'm Class A—Director Class. I can go directly to the Full Directorate. I ought to be able to hold them off, keep back the euth indefinitely. The odds are a billion to one against us if we try to break past—"

She broke off.

"But you don't gamble," she continued slowly. "You don't go by odds. You *know* what's coming. You've seen the cards already." She studied his face intently. "No, you can't be cold-decked. It wouldn't be possible."

For a moment she stood deep in thought. Then with a quick, decisive motion, she snatched up the cloak and slipped it around her bare shoulders. She fastened the heavy belt, bent down and got her shoes from under the bed, snatched up her purse, and hurried to the door.

"Come on," she said. She was breathing quickly, cheeks flushed. "Let's go. While there are still a number of exits to choose from. My car is parked outside, in the lot at the side of the building. We can get to my place in an hour. I have a winter home in Argentina. If worse comes to worst we can fly there. It's in the back country, away from the cities. Jungle and swamps. Cut-off from almost everything." Eagerly she started to open the door.

Cris reached out and stopped her. Gently, patiently, he moved in front of her.

He waited a long time, body rigid. Then he turned the knob and stepped boldly out into the corridor.

The corridor was empty. No one was in sight. Anita caught a faint glimpse, the back of a guard hurrying off. If they had come out a second earlier—

Cris started down the corridor. She ran after him. He moved rapidly, effortlessly. The girl had trouble keeping up with him. He seemed to know exactly where to go. Off to the right, down a side hall, a supply passage. Onto an ascent freight-lift. They rose, then abruptly halted.

Cris waited again. Presently he slid the door back and moved out of the lift. Anita followed nervously. She could hear sounds: guns and men, very close.

They were near an exit. A double line of guards stood directly ahead. Twenty men, a solid wall—and a massive heavy-duty robot gun in the center. The men were alert, faces strained and tense. Watching wide-eyed, guns gripped tight. A Civil Police officer was in charge.

"We'll never get past," Anita gasped. "We wouldn't get ten feet." She pulled back. "They'll—"

Cris took her by the arm and continued calmly forward. Blind terror leaped inside her. She fought wildly to get away, but his fingers were like steel. She couldn't pry them loose. Quietly, irresistibly, the great golden creature drew her along beside him toward the double line of guards.

"There he is!" Guns went up. Men leaped into action. The barrel of the robot cannon swung around. *"Get him!"*

Anita was paralyzed. She sagged against the powerful body beside her, tugged along helplessly by his inflexible grasp. The lines of guards came nearer, a sheer wall of guns. Anita fought to control her terror. She stumbled, half-fell. Cris supported her effortlessly. She scratched, fought at him, struggled to get loose—

"Don't shoot!" she screamed.

Guns wavered uncertainly. "Who is she?" The guards were moving around, trying to get a sight on Cris without including her. "Who's he got there?"

One of them saw the stripe on her sleeve. Red and black. Director Class. Top-level.

"She's Class A." Shocked, the guards retreated. "Miss, get out of the way!"

Anita found her voice. "Don't shoot. He's—in my custody. You understand? I'm taking him out."

The wall of guards moved back nervously. "No one's supposed to pass. Director Wisdom gave orders—"

"I'm not subject to Wisdom's authority." She managed to edge her voice with a harsh crispness. "Get out of the way. I'm taking him to the Semantics Agency."

For a moment nothing happened. There was no reaction. Then slowly, uncertainly, one guard stepped aside.

Cris moved. A blur of speed, away from Anita, past the confused guards, through the breach in the line, out the exit, and onto the street. Bursts of energy flashed wildly after him. Shouting guards milled out. Anita was left behind, forgotten. The guards, the heavy-duty gun, were pouring out into the early morning darkness. Sirens wailed. Patrol cars roared into life.

Anita stood dazed, confused, leaning against the wall, trying to get her breath.

He was gone. He had left her. Good God—what had she done? She shook her head, bewildered, her face buried in her hands. She had been hypnotized. She had lost her will, her common sense. Her reason! The animal, the great golden beast, had tricked her. Taken advantage of her. And now he was gone, escaped into the night.

Miserable, agonized tears trickled through her clenched fingers. She rubbed at them futilely; but they kept on coming.

"He's gone," Baines said. "We'll never get him, now. He's probably a million miles from here."

Anita sat huddled in the corner, her face to the wall. A little bent heap, broken and wretched.

Wisdom paced back and forth. "But where can he go? Where can he hide? Nobody'll hide him! Everybody knows the law about deeves!"

"He's lived out in the woods most of his life. He'll hunt—that's what he's always done. They wondered what he was up to, off by himself. He was catching game and sleeping under trees." Baines laughed harshly. "And the first woman he meets will be glad to hide him—as *she* was." He indicated Anita with a jerk of his thumb.

"So all that gold, that mane, that god-like stance, was *for* something. Not just ornament." Wisdom's thick lips twisted. "He doesn't have just one faculty—he has two. One is new, the newest thing in survival method. The other is old as life." He stopped pacing to glare at the huddled shape in the corner. "Plumage. Bright feathers, combs for the rooster, swans, birds, bright scales for the fish. Gleaming pelts and manes for the animals. An animal isn't necessarily *bestial*. Lions aren't bestial. Or tigers. Or any of the big cats. They're anything but bestial."

"He'll never have to worry," Baines said. "He'll get by—as long as human women exist to take care of him. And since he can see ahead, into the future, he already knows he's sexually irresistible to human females."

"We'll get him," Wisdom muttered. "I've had the Government declare an emergency. Military and Civil Police will be looking for him. Armies of men—a whole planet of experts, the most advanced machines and equipment. We'll flush him, sooner or later."

"By that time it won't make any difference," Baines said. He put his hand on Anita's shoulder and patted her ironically. "You'll have company, sweetheart. You won't be the only one. You're just the first of a long procession."

"Thanks," Anita grated.

"The oldest survival method and the newest. Combined to form one perfectly adapted animal. How the hell are we going to stop him? We can put *you* through a sterilization tank—but we can't pick them all up, all the women he meets along the way. And if we miss one we're finished."

"We'll have to keep trying," Wisdom said. "Round up as many as we can. Before they can spawn." Faint hope glinted in his tired, sagging face. "Maybe his characteristics are recessive. Maybe ours will cancel his out."

"I wouldn't lay any money on that," Baines said. "I think I know already which of the two strains is going to turn up dominant." He grinned wryly. "I mean, I'm making a good *guess*. It won't be us."

The Morphology of the *Kirkham* Wreck

Hilbert Schenck

The Riches of the Commonwealth
Are free strong minds, and hearts of health;
And more to her than gold or grain,
The cunning hand and cultured brain.
—Robert B. Thomas, *The Old Farmers Almanack,* 1892, William Ware and Company, Boston, Mass.

When the three-masted schooner *H. P. Kirkham* stranded on Rose and Crown Shoal southeast of Nantucket Island on January 19, 1892, the Coskata Life Saving Crew, led by Keeper Walter Chase, responded. The ensuing rescue attempt involved alterations in the local time flow of magnitudes never before observed within this continuum. Evolutionary physical forces were changed beyond the control of time-using peoples, and a fundamental question was introduced into the information matrix of this continuum, having, apparently, no resolution.

Time-using societies had always recognized the possibility that energy-users might attain significant mastery of time manipulation. Indeed, even occasional members of Keeper Chase's world group had, under the impetus of some violent or emotional event, been able to perform some limited and simple feats of time engineering, usually associated with mood and incentive control of others in the immediate situation. What became evident when the *Kirkham* stranded was that extreme-value probability theory could not set a limit on such activity by an energy-user totally motivated and having what Keeper Chase's peoples would incorrectly call a high level of "psychic" ability but what in fact is simply the ability to make information transfers within an altered time domain.

The northern gale blew shrieking along the back of Great Point, driving the spume off the wave tops and over the bitter beach. The patrolman crouched behind a sand hill, hunched to keep an occasional swirl of snow out of his collar, staring dully out at the white and gray sea. The wind had built up through the night, and now the shreds of dawn were blowing south over Nantucket, and the wind spoke continuously of urgent death.

The beach patrol, a hulking dark figure, turned to put the blast behind him and started back toward the station, where watchers in the cupola could relieve him in the light of day. It was twelve degrees above zero, with the wind gusting over forty.

Inside the Coskata Station the dark, shadowed paneling glowed faintly pink, reflecting the luminous brilliance of a huge coal stove in the center of the big common room. Nyman was cookie that week, and the wheatcakes were piling up on the cookstove in the small galley under the stairs. Four men sat silent, waiting for their breakfast, not trying to speak against the whines and rattles of the wind gusts. Yet they clearly heard the telephone tinkle in the cupola. A moment later, Surfman Eldridge appeared at the top of the stairs. "Skipper? It's Joe Remsen at Sankaty Light."

Keeper Walter Chase rose in the dark glow of the station, a giant, almost seven feet tall; his huge shadow startlingly flew up to obscure the walls and ceiling as he moved in front of the ruddy stove and up the stairs.

And Surfman Perkins, toying with his coffee mug, listening to the wind snapping and keening around the station, knowing that dawn calls from the lighthouse meant only one thing, suddenly realized for the first time in his life that he might die. He coughed, sharp barks of sound contrasting with the heavy, measured tread of Keeper Chase mounting the two flights to the cupola.

"Walter Chase here. Is that you, Joe?"

"Walter!" An urgent tone. Chase sensed that time was beginning to run away from him. "Masts on Bass Rip. We saw a flare last night late, but couldn't tell where. She's leaning some. Seems steady, but it's awful far to tell."

"What's her true bearing, Joe?"

"Just about due east from us. That would put her on the north end of Bass Rip."

Keeper Chase consulted a chart and compared angles. He looked out over the station pointer with powerful glasses. "Joe, I can't make her out. She has to be further out. We've got forty feet here and I could sure see her if she was laying on Bass Rip. She's got to be on Rose and Crown: South end from your bearing."

A pause. "Well . . . I don't know, Walter. I doubt we'd see her so clear that far. She may have lost her topmasts."

There was no point in arguing. Chase knew the wreck was fifteen miles out, on Rose and Crown Shoal. A sudden gust blew through the stout government sashes and swirled its chill into the cupola. The little tower rattled and shook. Walter Chase looked out at the ragged dawn, across at Eldridge, then down at the phone. "Joe, hang on. I'm getting the surfboat ready. We'll haul to the backside and launch there. I'll be back to you before I leave the station." Chase rose, ducking his head instinctively in the small room, and slowly climbed down the ladder, his mind fragmenting, working the launch, estimating the tide rips, laying beside the stranded vessel. "Eat quick!" shouted Walter Chase down the stairs. "We got a wreck on Rose and Crown!"

The difficulty in predicting improbable, time-controlling events by energy-users stems from these people's unlikely and illogical motivations and perception. One might assume that Keeper Chase's need to "defeat" the seas of the Nantucket South Shoals flowed from some sense of vengeance or hatred on his part resulting from the loss of a loved parent or a woman in some sea disaster. In fact, Keeper Chase suffered no such loss. Distant family members had, through the years, died on various whaling and trading voyages, but they were only names to Keeper Chase with little emotional attachment. Yet where the winter storms easily broke and ruined other capable men, for Keeper Chase the natural variation of wind and sea, so implacable and daunting to most of Chase's world group, only resonated with his self-image. In essence, Chase did not strongly believe in the "God" concepts so typical of energy users, but he strongly believed in a "Devil"—that is, the continual temptation of his world group by easy choices and safe paths. Keeper Chase saw the variability of the ocean as a natural test of behavior, as a kind of "Devil's assistant." That this naïve motivation coupled with his great physical strength and the urgent and marginal situation at the stranded *Kirkham* should have produced such an unprecedented control over time flow cannot now be understood. Keeper Chase's meaning and purpose in this continuum thus remains inexplicable, as in fact he himself was to realize.

Surfman Flood ducked around the corner of the station, finally relieved of the wind blast at his back. He saw the stable door was open, and in the dim interior, Perkins and Gould were fastening a long wooden yoke across the neck of the silent ox. Harness bells tinkled, sound pinpoints in the rush and scream of the wind. Flood's heart seemed hollow. "Where's it at?" he asked at the door.

Josiah Gould peered from under his slicker hat. "Rose and Crown."

Flood sighed. "Fifteen miles downwind."

"Ayeh. Better get some breakfast."

Flood pushed open the station door and felt the relative warmth and stillness of the dark interior suck at his resolve. Nyman was steadily lifting forkfuls of flapjack into his mouth, alternating with steaming coffee from a huge cup in his left hand. Across the table was Flood's place set with a heaping meal.

"We got some rowin' to do, George. Better feed your face

quick," said John Nyman. As they ate, rapidly and silently, the two wide doors of the apparatus room opened on the other side of the station, and swirls and draughts of chill rushed everywhere. They heard the shouts and tinkles as the stolid ox was backed over the sills and the harness lashings connected to the surfboat cart.

"Gawd, John, hits just awful on the beach!"

Nyman grinned and winked in the dark, chilling room. "The govinmint only pays you to go out, George," he said quoting the old wheeze. "You got to get back any way you can."

"Fifteen miles to windward! Hell's delight, we won't row a hundred yards in this smother!"

They heard Keeper Chase's deep voice in the apparatus room as the creak of the wheels signaled the surfboat's movement out into the wild dawn. He came into the common room and looked at the two men. "You fellers follow across the neck when you're finished and bring back the ox. I'm going to call Joe Remsen at Sankaty and have him order a tug from Woods Hole. We hain't going to row very far in this blow after we get them fellers off the wreck."

"Amen," said Surfman Flood under his breath. The wind was penetrating everywhere in the station, and the commotion was restless and insistent.

Walter Chase climbed back up into the cupola and cranked the phone magneto.

"Sankaty Light, Keeper Remsen."

"Joe, Walter Chase again. We'll be launching pretty quick. Can you still see her out there?"

"Hang on . . . yep. No change in her heel, as far as I can tell."

"Joe, will you call the town and have them telegraph Woods Hole for a tug. I think this storm's got another day to run, and we just hain't going to row back against it."

"Walter, I'll do my best . . . Them salvage fellows . . . they're hardly what you'd call heroes, you know."

Walter Chase grinned in the dark tower, which was suddenly shaking like a wet terrier. "Rats, vultures, buzzards, and skunks is what I usually hear them called, Joe. But we'll get back. Listen, Joe, I'm taking a line and drail. Might be some squeteague in those shallows in this rough weather."

But Joe Remsen made no sudden answer. He had rowed in the surfboat with Walter Chase under old Captain Pease when the Coskata Station had opened eight years before. Together they had worked the wreck of the infamous brig *Merriwa,* manned by a crew of New York City thugs who attempted to shoot up the station soon after they were landed. Walter Chase and an ax handle had secured the pistols, and then he and Joe had gone with them, now drunk as lords, to town in Wallace Adams' catboat. And Joe Remsen, feeling the tough and solid tower of Sankaty Light vibrate as a thin scream of icy air pierced the solid masonry, smiled in spite of himself, remembering the lunch at the American House. One of the drunken hoodlums had shoved his hand under a waitress's dress, and she had let him have a full tray of food plumb in the face. Back to back, he and Walter Chase had fought the six of them, chairs flying, crockery smashing everywhere. Joe Remsen's throat had a catch. He had to say something. "Walter . . . old friend, take care . . . God bless."

The walk across the neck took only a few blustery minutes, and Walter Chase met Nyman and Flood midway in that walk leading the ox home. Chase strode through the tidal cut between two high dunes, and the full wind caught his slicker and blew it suddenly open so that for a moment he seemed impossibly huge in the gray, fitful light. The surfboat lay above high water, and the men around it huddled together, their backs against the cutting wind.

Surfman Jesse Eldridge was number one in the Coskata crew. He walked, hunched and stolid, to Walter Chase. "It's going to be a tough launch, Skipper. Them waves are running almost along the beach," he shouted.

Chase nodded. "We're getting some lee from Point Rip, Jesse, but we'll have to launch across them, hold her head to the east. Otherwise, we'll be back ashore before we know it." They watched the breaking curls running toward them from the north.

Nyman and Flood came back, and the men, three on a side,

began to shift the surfboat into the backwash. The wind blustered at them. "We got to go quick . . . when we go, boys!" shouted Walter Chase.

The blow was slightly west of north, but the waves were running directly south and meeting the beach at a sharp angle. "Take her out about nor'east!" shouted Walter Chase. "Ready. . . . Now, jump to it!"

The six men lifted the boat by its gunwales and ran into the waves. A large group had passed and now the nearshore was a confused and choppy mess. The leading surfmen, Cathcart and Perkins, were almost up to their waists, and over the sides they vaulted, lifting and dropping their oars in the rowlocks. Now Gould and Nyman scrambled in, then Eldridge and Flood. Walter Chase pushed the surfboat out alone, deeper in, and now a curl appeared more from the east than the others and slapped the surfboat's bow to port back towards shore. Walter Chase moved his right hand forward along the starboard gunwale and pulled sharply. The twenty-three-foot boat gave a hop and her bow shifted eastward again. Then Chase was gracefully over the stern and the men were rowing strongly while he put out the long steering oar. They were clear of the shore break and moving into deeper water. Yet even here the waves were huge, rolling by under the boat and now and again breaking unexpectedly under the keel or beside them as they pulled together.

George Flood, cheerful and round-faced, was rowing port oar next to Jesse Eldridge. "Say, Skipper," he shouted up at Walter Chase. "I'm sure glad your ma never stinted you food. We must have been in a fathom of water before you climbed in."

Walter Chase thought a moment. "Actually, George, I hain't all that big, as Chases go," he boomed. "My great-uncle Reuben Chase was harpooner with Cap'n Grant on the *Niger*, and he went over seven feet. They claimed he could play a bull walrus or a whale on a harpoon line like you or I would a blue or striped bass." Chase paused, then . . . " 'Course, that would be a *small* whale, you understand."

Josiah Gould, seated directly ahead of Eldridge, lifted his head, his huge mustache blowing every which way. "Hain't that awful!" he yelled. "He's not just taking us out here to catch our death from pee-nu-monia, but now we're going to listen to more of them Chase family lies, too!"

Flood grinned over his shoulder and shouted back. "Them's not exactly lies, Josiah. Them's what's called 'artistic license.' "

Walter Chase looked benignly at Flood, his small eyes bright and his sideburns wild and full in the whipping wind. "I wisht I had your education, George. It's a plain wonder how you fellers with schooling can call one single thing by so many names. Now my daddy always said there was just three kinds: plain lies; mean dirty, awful lies; and what's in the *Congressional Record*."

Gould and Nyman looked sideways at each other, winking. If they could get Skipper Chase going on them "govinment fellers," it would be a short and cheerful run to Rose and Crown.

But the wind was worse. They were completely clear of any lee from Great Point. Even Chase's huge voice would be torn away and mutilated. "We're . . . far enough . . . out! Get . . . the sail up!" The four stern oarsmen continued to row, now more northerly into the teeth of the blasting wind. Cathcart and Perkins brought their oars inboard and wrestled the sail, tied in a tight bundle, out from under the thwarts and up into the wind. With Gould's and Nyman's help they finally stepped the mast and then unfurled and dropped the small lateen rig. It caught and filled with a snap, and Walter Chase wrenched the steering oar so hard to port that it described a long arc between the water and the steering notch in the transom. The boat darted off. Eldridge manned the sheet, and the other men huddled on the floorboards, their heads hunched inside the thick issue sweaters and stiff slickers.

Chase, at the steering oar, and Eldridge, on the stern thwart, had their heads close together; and now, running with the wind on the stern quarter, they could suddenly speak less stridently.

"I'm going to head for the lightship south of Great Round Shoal, Jesse," said Walter Chase, his arms in constant motion. "If it comes on to blow worse, we'll just have to go on board her. If we decide to keep going, we'll lay off sou'east and run down to Rose and Crown."

Eldridge was silent, then: "When do you figure we decide, Skipper?"

Chase looked at the jagged seas, whitecaps everywhere to the horizon. "Much beyond the Bass Rip line, we could never fetch the lightship. This lugger hain't much to windward."

The two men looked out ahead as the surfboat, heavily driven, wallowed and yawed and fought the pull of Walter Chase and his tough hickory steering oar.

Now they were three miles out, and looking south, Walter Chase saw that Bass Rip was clear and that the wreck was certainly on Rose and Crown. "We got to decide, Jesse," said Walter Chase.

Surfman Eldridge looked down at his high gum-rubber boots and nodded. "It hain't got worse, Skipper."

Chase's small eyes glittered. "Let out the sail, Jesse," he said, and the surfboat bounced and slapped and rolled, but now it was better, for the waves were astern. Off they dashed southeast, surfing down the long rollers in the deep water, then struggling up the shifting water hills. Between the Bass Rip line and McBlair's Shoal, Walter Chase first saw the masts of the wreck. She was at the south end of Rose and Crown, probably in that one-fathom spot there, and leaning to the south perhaps twenty degrees. He headed a bit more southerly and they left the choppy white smother of McBlair's Shoal behind.

Now the three masts were clearly visible. The vessel lay roughly east and west with her stern to Nantucket. She had struck and then bilged, and now the waves were breaking cleanly over her. They had driven the hull over to starboard so that a spectacular line of surf would suddenly appear all along her port side that canted up to face the seas. They were too far away to count the men, but Chase could see dark forms in the ratlines. He peered intently at the wreck. Was it shifting now? It was a bad stranding! If she were facing the seas, even quartering, but broadside they were wrenching her. And the tide was coming. The seas would enlarge and she'd be hit even harder. Chase peered and peered at the wreck, and the surfboat drove along the line made by his eyes.

* * *

Chase Two was aware of the *H. P. Kirkham* in a total sense. She was not going to stay together any longer. Chase Two detected unbalanced forces within the ship-sand-wind-water field matrix. He penetrated the force structure around the *Kirkham,* but there was no TIME! The surfboat was running down the seas. The *Kirkham* was twisting as the combers, steepened and shortened by the shoal, boarded her with shuddering blows. Chase Two clinched, and time flowed more slowly. The waves moved like molasses. The shocks were stretched out, and he could trace the force imbalances. SLOWER! He could not speed the finite duration of impulses flowing between his billions of neural cells, but he could slow time and process data that way. Fiercely he clinched! Time, he realized, could be traded for information. He saw the *Kirkham* completely, and yet simultaneously in every relevant detail. The mizzenmast was shaky, split. Not much had shown on the mast's surface, but now the stick was resonating with the wind, and the splitting was worse. It would soon bring down the main and foremast, and the men as well.

There was no solution within the energy matrix alone, and neither time nor information domains extended directly into the energy system. The mast would have to be replaced.

Chase Two stooped like a hawk down the *Kirkham's* time line. He saw her leave Rose and Crown Shoal and flow backwards to Halifax and leave her lading. Then faster, backwards to other voyages in her brief year of life. Now the masts were out and the hull was coming apart on the stocks of a boatyard near the tiny town of Liverpool on the south coast of Nova Scotia. The masts suddenly grew branches, and in a twinkling, Chase Two watched a French timber cruiser looking up at a tall pine deep in the Nova Scotia forests. The cruiser turned to his associate, the shipyard boss's young son on his first wood-buying trip into the woods. "By gar, dat's one fine tree, eh?"

The young man nodded. It was the tallest in the area. But now Chase Two showed the Frenchman something he had not seen before, that other time. The tree had been struck by lightning. The scar was grown over, but you could just make it out curling from the top and disappearing around the trunk.

"Look," said the timber cruiser. "Dat tree been struck. We walk around." And on the other side they saw the faint scar traveling down to the ground. "Risky, dose ones," said the Frenchman in a superior way. "Hmmmmm." And as he looked around, Chase Two showed him a shorter but perfectly branched mast tree on the other side of the clearing, and the French cruiser pointed and smiled. "Not so beeg, dat one, but plenty tough, I teenk."

And back down the time line, Chase Two dropped like a stone. He saw the new mizzen erected on the Liverpool ways, then, faster and faster, the loading and unloading and movement until the *Kirkham* again struck on Rose and Crown Shoal, bilged, and lay through the stormy night with her men in her rigging.

When Keeper Chase learned that informational and temporal entropy flows could be interchanged, his power to influence events grew at an unprecedented pace. In the course of replacing the *Kirkham's* mast, Keeper Chase solved a variety of hydrodynamic and structural problems of extreme complexity and entirely by inspection and processing of data. Much more significant, he dealt surely with the philosophical and practical problems of time-information interchange and realized that if time flow could be slowed, it could be controlled in other ways. His ability to arrest time flow within his local region was now so pronounced that a detectable chronologic entropy gradient existed within the entire continuum.

The surfboat blew down on the stricken schooner from the north, heading directly for her battered port side, where white spume flew up twenty feet or more when a big wave took her full on.

"You bow men," shouted Walter Chase. "Get the anchor ready." The positioning had to be done correctly the first time. There would be no clawing back up from the schooner's lee to reanchor if they did it badly. Walter Chase watched the choppy, surging space shorten between the surfboat and the schooner. The current was running to the northeast with the wind a bit west of north and the waves about from due north. He decided to anchor upwind of the vessel's stern and then lay back south and easterly to come under the mainmast and her center ratlines, where the crew was now clustered.

Chase's small eyes gleamed in the gray, dull light. He watched the distance shorten and the schooner widen and her masts grow up and, in them, the men now clearly seen.

"Watch your head, Jesse!" shouted Walter Chase. "We're rounding up now!" He put the steering oar hard over, and again it formed a bow of iron-hard hickory, arched against the forces that drove the boat halfway around, heeling and wallowing wildly until it faced the screaming wind and sharp seas. "Anchor over!" shouted Walter Chase; then, "Oars out, all of you!"

They were up on the thwarts holding her head against the wind as she slipped back with Perkins paying out the anchor line over a smooth, maple cleat. The surfboat lay on her tether about southeast, and Walter Chase guided her back and back until they were a few yards from the schooner and just beyond where the big rollers broke and shuddered the vessel all along her length. The surge was ten feet or more. The surfboat lay down in a trough, and they could look up and see several feet of the schooner's side, then up until they were above the rail and a great wave was sliding out from under them and creaming white and lovely over the vessel's port rail in a burst of foam and a sound of roaring and groaning that made Walter Chase flinch his cheek muscles, for he knew how weak the schooner was.

Chase cupped his hands and bellowed directly into the wind. "Perkins, throw them the heaving stick."

Perkins heard and readied the stick and its loops of line. The surge picked the surfboat up, and as they came level with the schooner's rail, Perkins hurled the stick into the rigging, with the thin line paying smoothly out behind in a graceful arc. One of the crew crawled up the ratlines to where the stick was entangled in the shrouds and turned toward the surfboat.

"You . . . bend a line on that! Use your topsail clew line."

The roar of Walter Chase's voice flew downwind, and in moments, the clew line was fast to the stick, and back it came, hand by hand, through the smother to Perkins, who bent it on the same cleat as the anchor warp.

The other end of the clew line was in the hands of the sailor and two others who had crawled over to join him. Walter Chase shouted again. "Tie that line to a shroud, you men!"

They stared stupidly at him, and sudden spray flew up in their faces. Walter Chase turned back to Perkins. "Start to haul in on that slowly. You rowers, ease us toward her side."

But the schooner's crew had waited long enough, and they, or three of them, began to pull fiercely on the clew line themselves. Walter Chase felt the boat jerk roughly toward the schooner and begin a deep roll broadside in a trough. He crammed the steering oar violently over and spun around, pointing at the schooner. "Stop hauling! Stop, I say! Make your end fast. If you make one more pull on that line, we'll cut it!" And as he spoke, Chase pulled a big clasp knife from his slicker pocket and opened the blade with a snick that pierced the duller voice of the gale. Then he passed the big silver knife forward to Perkins, who brandished it above the clew line. The men on the schooner saw the great dark figure with the knife and heard the huge voice driven down by the wind, and they tied off the line and huddled, dully watching the Coskata crew, using both rope and oars, begin to move toward the wreck.

Suddenly the schooner shuddered and inexplicably rolled to windward. She came almost upright and then went back over to starboard, stopping her breathtaking swing at the same list as before. The mizzen gaff snapped off and fell, thudding against its mast on the way down. The schooner began another roll to port, and Perkins looked directly at the men in the ratlines, and his eyes and theirs met. He remembered a Sunday six years ago when, after church, he and his mother had driven in the wagon over to Little Mioxes Pond where, everyone said, a large vessel had blown ashore. They had spent the day with hundreds of others watching the men in the rigging, too weak to grasp the lines shot over the vessel by the surfside crew, falling one after another into the raging sea. The vessel had stranded well out so the crew were only small black figures and they did not move very much when they fell, but Perkins never after that time shot another crow or grackle with the .22 Winchester pump that his mother had saved for a year to buy him. Even if they were just birds, they fell the same way, black against the far sky. And now these men were about to fall, blackly still, but he would see their eyes clearly this time.

"Gawd help us, Skipper! She'll shake her sticks out!" shouted Perkins in a choking, coughing voice, strident with terror.

Walter Chase had followed that roll with bright, keen eyes. She could not withstand much more of that! "No!" he said sharply.

Chase Three surveyed the flow field under the wreck and processed the observations. He clamped intensely on the time flow, and the *Kirkham* was motionless in a sea of stationary fluid and a sky of stationary wind. He explored the flow characteristics of the near shores in every particular, considering the special character of the *Kirkham's* fields of forces. The current, shifting clockwise during the flood as it did in the area, had undermined the sand bed on which the *Kirkham* lay. But worse, the current, now running more and more counter to the wind, was moving the hull as well.

Chase Three considered how the force and energy relationships could be corrected. The wind was beyond manipulation, deriving as it did from such a disparate mass of variables as to make significant time-based alteration impractical. But the flows of gravity and wind-driven water were another matter. As Chase Three studied these fields of flow, he gradually realized that the natural relationships allowed a bifurcation within the viscosity functions. There were at least two flow-field configurations that had equal probability, and most important, either could exist with no change in total energy level within the continuum. The present field system allowed a strong easterly current to move in over the shoal against the *Kirkham*, but with the alternate field the flow would be slightly damped and diverted more northerly, and the wind force on the schooner would be sufficient to hold her against the sand and damp the roll forces.

Since there was no energy gradient involved, Chase Three immediately altered the continuum to the new flow field, this information gradient being offset by the altered time flow in the local area. The *Kirkham* shuddered but did not roll again.

With his introduction of the Chase Field into the information matrix of our continuum, Keeper Chase was reaching the peak of his astonishing powers. That any alternate description of the fluid-dynamic field existed was not even known, and that Keeper Chase should have found a solution at equal energies was quite marvelous. He did not, when utilizing time-information entropy balances to make the shift, consider that these same laws govern the development and evolution of galactic and supra-galactic motion and that the field shift must occur there as well. Thus Keeper Chase, in addition to sustaining an extraordinary temporal gradient within the continuum, had now inadvertently but irrevocably altered the way in which the energy universe would develop. Those time-using peoples who existed outside the gradient now convened and considered the immediate situation. We too could work within altered time, but the randomness of what was occurring put us beyond normal information transfer procedures. The storm on the Nan tucket South Shoals had spawned a gradient storm in time itself. If the rescue attempt should become unlikely within any statistically allowed alternate energy structure, we would have to consider Keeper Chase's reaction to that perception and what an impossible but certainly powerful reaction by Keeper Chase to breach the energy-time-information barriers would cause within these boundaries.

Perkins, bent in a fit of coughing, saw that the schooner was stationary again. "Cathcart," shouted Walter Chase. "Bend a bowline in our painter and get ready to heave it over."

Now they edged closer, hanging like a lunging pony against the whipping anchor line. "Throw!" shouted Chase and the line flew across. "One man at a time," shouted Chase down the screaming wind to the schooner. "Put that bight around your waist."

A large, hulking Negro who had caught the painter passed it to a smaller figure, evidently the cabin boy. The youngster put the line over his head and waited, staring frozenly into the wind.

"Now . . ." and Chase's voice boomed under and around the wind's cry. "When I say jump, you come! You hear!"

The boy nodded, staring out at the marching lines of water foaming towards them.

"Cathcart, haul us in a bit . . . Now, steady, boys!" The surfboat was caught by a comber and lifted, up and up, and the wave was pushing the boat toward the schooner. They were on the peak and the curl was slipping past.

"JUMP!"

The boy flung himself off the ratlines, his legs flailing. Cathcart and Perkins handed in the painter as he fell, thudding, into the space between the bow and center thwarts. "Ease that bow line quick!" shouted Walter Chase, and the surfboat lay off to the east before an early break could turn them over at the schooner's rail.

The boy looked up from the floorboards, his ankle hurting, his teeth chattering; and over him loomed a gigantic figure, sideburns wild and blustery, eyes small and intensely bright, beacons against the wild gray sky. "We count six more, son. Have you lost anyone yet?"

Somehow the boy was able to speak. "No, sir. Seven in all. The cap'n—the cap'n ain't so well. We—we been in the rigging since eight last night. Gawd, it's . . ."

"What ship?" asked George Flood, turning suddenly around.

"*H. P. Kirkham*," said the boy. "From Halifax with fish. Bound to New York."

Walter Chase stood up. "Let's get the next one. You starboard rowers, bring us in slow. Cathcart, bring her head in."

Each time, they approached the *Kirkham* and waited for the proper wave to lift them up and slide the boat close. Then a black tumbling figure would come down into the surfboat every which way, limp with fear and exhaustion and dazed by the sudden, unexpected hope.

Now there were three left and the surfboat rolled more heavily and more water slopped over the gunwales. "Sir," shouted the cabin boy. "I think the cap'n's coming next. They're going to have to sort o' throw him."

Walter Chase peered at the three figures in the ratlines. The wind had slackened a bit and it seemed brighter. He could see an old man, conscious but unable to hold his head up, supported and held against the ratlines by a huge Negro and another big man in bulky clothes. "Josiah, Johnny . . . get ready to help when this fellow comes across."

The center rowers shipped oars and waited. Cathcart carefully pulled them in, a bit at a time. Then he cleated the clew line and hurled the painter back across the foamy gap. They put the loop over the old man's head and shoulders. The boat was rising. "Get ready!" shouted Walter Chase to the three men. "Now!"

The two men threw the captain feet first into the boat. He came down crossways, catching John Nyman across the cheek with his fist as he fell. His head thumped a thwart and he slumped, a bundle of rags, into the bottom of the boat.

Walter Chase quickly knelt and lifted the old man's head. "Keeper Walter Chase. Coskata Life Saving Station. Can you understand me, Captain?"

The old man, his whiskers white with frost and brown with frozen tobacco juice and spittle, stared back unseeing. "Aye. Captain McCloud, master, *H.P. Kirkham* out of Nova Scotia. Thank God . . ."

Chase's eyes pierced the old man's own eyes, and he nodded. "Captain McCloud, we cannot save your vessel. She is breaking up and this storm will grow worse by nightfall."

"I know," said McCloud and his head fell forward and his eyes shut and he shivered in cold and pain and despair. Then . . . "This bloody, foul, awful coast!" His eyes briefly lost their dullness. "Worse than Scotland! Worse than the Channel! These rotting shoals stick out so bleeding far . . . God Almighty . . ." The effort exhausted him. He did not speak again.

The next man was the first mate, hard, grizzled; Cockney-tough enough to sit up after his jump and stare at the young, slender Perkins, bent over in a fit of coughing. "Well, you blokes don't look like bloody much, but you bloomin' well know your business out here!"

And on their final surge up over the schooner's rail the huge black crewman flew between the great and little boats with a sudden grace, and he, like the mate, sat up immediately and peered about from huge white eyes. But he said nothing.

"Now, lads," boomed Walter Chase, as the surfboat lay off easterly, bobbing and pitching in the smother like a logy cork. "Oars out. We got to clear this shallows afore the wind comes on. Lively now."

The four stern oarsmen pulled mightily while Perkins and Cathcart heaved on the anchor warp. Slowly they moved to windward, their efforts sending rivers of sweat inside the heavy sweaters and slickers in the twelve-degree, forty-knot blast.

"Anchor up, Skipper!" shouted Cathcart while Perkins suddenly bent double, both hands over his mouth.

Walter Chase looked back, his side whiskers black spikes, his huge slicker masking the *Kirkham*. He hated to give the gale an inch, but to get past her stern to the west would be a near thing, and the wind was rising again. He put the steering oar over and they fell off on a big soft wave to starboard. "Pull, boys, we got to stay ahead of these combers."

They rowed eastward, then more southerly and cleared the *Kirkham's* smashed and sagging bow by forty feet. Walter Chase put his oar to starboard and they pulled under the schooner's lee. It was easier there. The schooner was acting as a breakwater, taking the big ones before they reached the surfboat, and they pulled strongly to the west, the wind hard and vicious on their starboard quarter and the sea confused and breaking everywhere, an endless mouth filled with shifting teeth.

But once beyond the schooner's length it was impossible. Chase put her more toward the south, taking the wind on the beam with the current still northeast and running those great, curling rips in the very shallow spots.

"Jesse," said Walter Chase, leaning forward. "We got to clear this shoal afore the high tide this afternoon. Them rollers'll start to break and we couldn't lay at anchor. And the wind's making up again. Them clouds are coming back."

Jesse Eldridge only nodded. He was pulling too hard to talk. They were moving southeast, but only barely. The *Kirkham* was close behind them, and the surfboat was slopping about, taking splash on every wave.

Walter Chase looked at the men they had rescued. His little bright eyes fixed on the first mate and the Negro, their heads buried in coats against the chill. "You fellers. Yank that sail and mast out of there and pitch it over the side. Our sailing days are done!"

The men moved slowly, as best they could, helped by a hand from this or that rower, and finally the outfit went over the side in a piecemeal fashion, trailing astern and finally pulling loose.

George Flood looked up and winked at Walter Chase. "Skipper," he panted, "how you going to explain throwing that valuable govinmint property over the side to the inspector?"

Walter Chase, at that moment fighting a great, half-breaking wave that threatened to broach the surfboat, suddenly winked a gleaming eye back. "George, I'll just tell that feller that we met this here bureaucrat adrift on his very own desk looking for Washington, D.C., and we just plumb did the Christian thing and loaned him our sail."

Charles Cathcart, leaning intently forward as he pulled, burst into a roar of laughter. "Hell's fire, Skipper! They'd just say you didn't get him to fill in the right forms."

Perkins's oar trailed astern, and he leaned over the side, vomiting and coughing great, deep, sharp barks above the gale. Cathcart reached towards him, and the surfboat lost way and began to bounce and shift southerly into the troughs. Walter Chase looked piercingly at his men. They *must* clear the shoal now. It would only get worse.

Chase Four entered Perkins's continuity of self-awareness. The boy was sick, probably pneumonia, for his lungs were very wet. He was beaten. The cockney mate's praise had got him through the anchor recover, but now he was completely involved with his cough and nausea.

Chase Four dropped down Perkins's time line seeking a point that would reverberate with the *Kirkham* rescue. . . .

Each year on the last day of July along the islands, the life-saving crews return from a two-month off-duty period to a ten-month routine of patrols and watches. On that night the previous summer, the Coskata crew had produced their usual party. They had hired a banjo and violin from town, asked their wives, relatives, friends, and suppliers to the festive evening, and cleared out the dark apparatus room of its large gear. Colored streamers hung from the suspended life oar, and festoons of buoys made arches beneath which the dancers turned. The girls, slim and pretty in ankle-length dresses, puffed sleeves, and swinging hair ringlets, smiled at the tall men in their government blue. Before the light went, they trooped outside to the breeches buoy training tower, and the girls climbed, one above the other up the ladder, and all looked back smiling while George Flood pressed his Kodak button and gave a happy shout.

But the prettiest there was Abigail Coffin with Roland Perkins. When the others returned in the dusk to the laughter and screech of the fiddler's bow, he caught her arm. "Let's go look at the ocean, Abby," he said. She was the nicest girl in the town, always smiling, her eyes so bright and full; and as they walked away from the station, Perkins could barely breathe, his chest was so full of love and hope. "Abby . . . could I . . . would you . . . ?" and he leaned toward her and brought his other arm up behind her back.

Chase Four did not wait for the sharp and hurtful reaction; he had skimmed by it once. Instead, he showed Abby Coffin that Roland Perkins was actually a fine, handsome boy. As she looked at him, she realized how sensitive and brave he would always be, how good and gentle his thoughts were toward her. She turned her face upward and they kissed. Later, under a bright moon, she said breathlessly, "Yes, you can touch me there, Roland."

* * *

Perkins, his coughing fit mastered, nodded at Cathcart and began to row strongly. Walter Chase urged them on. "We got to make some depth, boys." Perkins, grinning to himself, pulled and pulled. He knew they would get these men back. Chase was too good a boatman to fail, whatever the wind. They would all get government medals. And he thought of Abby and the medal and how she would hold him when he told her. Slowly the surfboat left the *Kirkham* behind.

George Flood's eyes popped open. "Look back quick, Skipper!" he shouted. Walter Chase spun around. In that instant, the *Kirkham* was dissolving. Her foremast was halfway down, with her main following. The taut and snapping shrouds ripped the quarter-boards completely off the starboard side, and the deck buckled in several large pieces. The mizzen fell, and before it struck the water, the entire hull had disappeared. She had gone like smoke in a gale. Flood looked up at Walter Chase. "Dang lucky we didn't wait for another cup of coffee at the station, Skipper!" he shouted.

Walter Chase, fighting the steering oar continually in the heavy and confused seas, still stared back at the unchecked rollers now streaming over the *Kirkham's* last berth. They had taken the last man off less than an hour before. His eyes narrowed and he wondered about the rescue. Everything was so damn near, so chancy.

The Coskata crew rowed and rowed on Rose and Crown Shoal. Sometimes the boat moved west and sometimes it paused and pitched. Noon was past and the sky had darkened again. The wind was rising with the tide, but they were slowly getting into deeper water, into the twelve-fathom channel that cut aimlessly between Rose and Crown and Bass Rip. The waves were longer and not so steep, but the wind was too heavy. They were hardly moving and the men were exhausted. And the current had revolved almost due easterly and was actually setting them back away from Nantucket. This would have to do for now.

"Cathcart! We got to anchor. Handy now!" They lay back with the winds hammering their starboard quarter, all the scope they could muster laid out to their biggest anchor. The men slumped over their shipped oars while Walter Chase shoved the steering oar this way and that, using the current run to steer his boat up and over the combers. The wind was building again, and its scream and slash was icy and terrible. Jesse Eldridge, hunched in a nest of sweaters and slicker, looked up at Walter Chase. "Skipper, we didn't even make a mile in three hours. You think that tug'll get out here?"

"I figure he will as long as he thinks there might be some loot on the schooner, Jesse," said Walter Chase. He sensed, in fact, that the tug would not come into these wild shoals. Bitterly he thought of the wonderful strength of her cross-compound steam engine driving that big powerful screw. Yellow, rotten cowards! What was the point of even building such a vessel if you could not find men to man it? The surfboat jumped and tugged at the snapping anchor line, while the crew bailed as the spume and spray came in on them with every wave. Perkins and Cathcart gently tended the anchor line, wrapping it in rags, shifting it a few inches now and then to relieve the chafing.

The sky grew darker as the afternoon wore on and the wind built up again. There was so much agitation and violent activity, so many unexpected swoops and thumps, so many waves that appeared from odd directions and with surprising steepness.

Chase pulled and fought the oar, staring out at the screaming bowl of energy around him while a coldness and fierceness steadied his heart and mind. He would bring these men home, all of them. Nothing anywhere was more important than that. The tug, the lifesaving service, the men at their desks in Washington and Boston, the sea and its commerce, the life cars and motor-driven surfboats, the rescues of the past and future, men adrift on the seas of the world and foundering forever in the gales and currents along the coasts, meant nothing beside these few in the Coskata surfboat. He focused his great strength on this single purpose and found a balance between the forces of the storm and his own resolve. They pitched and waited in the freezing blast for the tide to turn.

Dusk comes early to Nantucket in January, and it was almost dark by the time the surfboat had swung clockwise on her tether and now lay a bit west of south. Chase knew they had to go whenever the tide could drive them, and he shouted and joshed the men. Tiredly they put out oars, pulled up the anchor, then struggled off to the west. Chase used his rowers to hold a northerly set, counting on the southeast current to give them a general westerly direction. They took plenty of slop with the waves on their starboard bow, and Chase urged the *Kirkham's* mate and her black crewman to bailing. The boat moved into deeper water, and as the night came on, Chase suddenly saw, on the very rim of his world, the tiny, flashing point of Sankaty Light.

"Hey, boys!" he shouted. "There's old Sankaty and Joe Remsen having fried bluefish for supper with a bit of Medford rum and lime in hot water."

"Dang me, Skipper," said George Flood, "I wouldn't mind the rum, but Joe can keep the bluefish."

"George," said Walter Chase shaking his head sadly, "I can't make out how you fellers can call yourselves Nantucketers when you like that awful, smelly cod better'n a little fried blue."

This discussion, which ebbed and flowed at the station depending on who was cook that week, somehow cheered Jesse Eldridge immensely. "Walter," he said, loudly and firmly, "even them rich Boston summer folk won't give a nickel a pound for blues. You know that as well as I do."

Chase leaned on the oar and turned them a bit more northerly, staring off at the lighthouse. He roared with laughter. "Jesse, them Boston folk smack their lips over three-day boiled cabbage and corn beef, flaked cod that would turn a hog's stomach, and fin and haddy so hard it would break a shark's jaw. Hell's delight, they wouldn't even *notice* a nice hot little blue laying in a nest of parsley, new potatoes, and melted butter."

They rowed on and on toward the light, and the men turned now and then to stare at the pinpoint, so bright and yet so tiny against the black swirl of wind. When they turned back to where the *Kirkham* had been, they saw answering bright and tiny spots in Walter Chase's eyes somehow reflecting and focusing Sankaty.

By ten that night the wind was blowing a three-quarter gale, and the current was rotating to the northeast. They could not go against it, and Walter Chase ordered the anchor down again. Now the wind, filled with a fitful snow, was bitter, and the men slumped against each other, their sweat drying coldly under their clothes, their heads nodding. Chase continually worked the steering oar, roused the men as they drifted off into frozen sleep, ordered the bow crewmen to watch the chafing of the line, and continually rotated his head seeking the great seas moving in the dark. They suddenly appeared as dim, faintly phosphorescent mountains that dashed out of and into the dark at terrifying speeds.

In the intense and shouting dark, the seas loomed huge and unsuspected. There was a wildness about them, a wholly random cruelty. The storm had blown for two days and unusual current motions had been set going. Walter Chase's head swiveled back and forth. He sensed the movement, the surge and backflow. The chill ate at his bones, but his own cold resolve was more arctic still.

Chase Five examined the lumpy and stationary sea. He then examined the rate of change of the water profiles. This was a deadly business! The circulation due to wind stress had rotated the current further than usual to the east. This set up a possible amplification with the flow between Bass Rip and McBlair's Shoal. There was a statistical possibility of one or more resonant occurrences that night! Yet they were still relatively unlikely. No! Chase Five clamped the time flow even tighter and increased the gradient. Within minutes a resonance would actually occur! The wave would build at the north end of Rose and Crown, receiving energy from the cross flow and a sudden wind gust stress. It would break in a mile-long line just north of them and reach them cresting at eighteen feet. The chance of their staying upright was one in three. The chance of their not swamping was . . . nil!

Chase Five, within the theoretical bounds set by entropy flow requirements, stopped time utterly. The continuum waited as Chase Five's neural interconnections achieved a higher level of synthesis. He saw a single possibility. If this resonant wave was unlikely enough . . . yes . . . that was it! Extreme value probability theory could be modified within the time domain, providing that no significantly less likely event was occurring at that instant in the energy continuum. He could lower the expectation and make the wave more unlikely without interaction within the energy domain. Furthermore, it was not just the wave itself that was unlikely, but the wave interpreted by Chase Five, himself a most unlikely event.

That was it! He changed the probabilities, and the wave, instead of building towards its terrifying height, received its new energies at slightly different times and . . . No!

The wave was suddenly building again! Chase Five sensed some other manipulation. Staggered, he clamped tightly on the time flow and asked his first question:

"Who?"

When Keeper Chase modified the laws of extreme value probability within the continuum, he forced us time-using observers to become participants in his struggle with the storm. While highly unlikely events occur infrequently, they exercise a hugely disproportionate effect on the evolution of the continuum. Just as a coastline on Keeper Chase's world will lie unchanged for a hundred years, to be altered drastically by a single unlikely storm lasting a few hours, so the improbable but possible events in the evolution of stellar and information systems often determine the long-term character of huge volumes of energy and temporal space. We could not, then, allow such essential probabilities to be manipulated at the whim of energy storms and energy-users. Thus we intervened and canceled the change. Keeper Chase detected us at once and asked his first question. We decided to answer him . . . almost totally.

Chase Five received the full brunt of the information dump. Like the sky falling in from every angle, the answer to his question flowed faster than thought into his mind. It was an implosion of data, a total, sudden awareness of the continuum, of time and energy and information and their interactions. Of worlds and stars, creatures and spaces, hidden truths and intricate insights.

Chase Five was staggered. He clamped on the time flow and tried to organize it all. Like a swimmer, thrown deeply into the dark blue of the deep ocean, he fought and rose toward the light of day, moving through a boundless mass of data. Yet what was happening? Why was he so deeply involved with these others? How did the *Kirkham,* one in ten thousand among such schooners, and these men, a few among millions, come to be at the center of all this? Chase Five assimilated the focal points of the continuum, but he did not yet understand himself or the nature of his adversaries. Clamping and clamping on the time stream, he desperately asked his second question:

"God?"

Irony, in the sense understood by those in Keeper Chase's world group, is not a normal component of time-using organizations and duties. Yet Keeper Chase's second question to us achieved the exact essence of that special quality. For if there was a single conscious entity within the entire continuum at the moment who qualified as "God," in the sense of Keeper Chase's question, it was Keeper Chase himself. We could not determine how large an information excess Keeper Chase could tolerate, but his confusion seemed to offer us an opportunity. We responded with the remaining information that we had withheld the first time: we showed Keeper Chase how the continuum was organized within its various aspects and, finally, the nature of consciousness within this organization and its relationship with the information, time, and energy aspects of the whole.

The second dump of information was not as extensive, but far more staggering. For Chase Five finally saw himself within the total continuum. He saw the circularity and hermetic nature of his activity at the *Kirkham,* the unlikely, really senseless character of the rescue and how unimportant, really meaningless were the men now barely alive in the wet and pitching boat. Good Lord, what was the point anyway? His control wavered and time began to slip. The wind moved back towards its own natural pace. The seas became more independent. . . .

Now wait! Chase Five, in his puzzlement and despair, still processed data. And suddenly he saw the fallacy, the problem with their attacks against him. He steadied and clamped time. Yes! Yes, of course! He was stronger! The circularity didn't matter! What mattered was *only* the event! Everything led to that. And the more *unlikely* it was, the more *essential* it became. Yes! He, Walter Chase, Keeper of the Coskata Life Saving Station, was exactly and completely his own justification. And now Chase Five struck back at them. Masterful in his total control of information, gigantic astride the interlaced worlds of energy and time, he stated his third and final question. But because he completely dominated the continuum in all its aspects, he no longer asked. For he knew with complete certainty that none of them could deny what he stated.

"I am central to the evolution of the continuum. My control and my improbability are proof of that!"

At once the growing wave received its various inputs in harmless and likely sequences and passed under the Coskata boat as a huge but almost unnoticed roller. And with that, the storm on the Nantucket South Shoals began to die. For it, like all storms, had to obey the laws of probabilities, and after two harsh days, it was moving off and softening as it went.

Walter Chase, the steering oar now inboard as the wind slackened, saw that dawn and the new tide were coming together. "All right, boys! This time we'll get there!" he shouted. In came the anchor and off they went, the great seas cresting no longer, the wind lessening, and the temperature rising as snow squalls came and went, gray against a dull dawn.

On and on they rowed, and Walter Chase now became aware that Perkins looked odd. His eyes were shining, liquid and bright, and his cheeks were much too red and also shining strangely. The boy rowed as strongly as any, but Chase watched him with more and more concern.

"Perkins," shouted Walter Chase, "see if that mate from the *Kirkham* can relieve you for a while."

But Perkins was thinking of Abby on the beach. She would probably be there when they came in, for her brother was in the surfside crew and he would have told her that they were out. "I'm okay, Skipper," he said in a voice that Chase could barely hear. Chase peered through the snow at the rowers. Perkins was very sick. Perkins must . . .

Chase Six realized that Perkins was dying. The boy's level of consciousness integration had slipped drastically. Desperate, Chase Six plummeted down Perkins' time line seeking solutions everywhere. But the lessening of the storm had sapped his abilities. He could no longer clamp on time or integrate his hard-won information to tasks like this. Yet his very agony gave him the control to achieve the data that crushed, and crushed again, his hopes. How tenuous and marvelous self-awareness was in the continuum! How delicate, beyond yet embedded within the energy system, linked with loops of information, operating within and yet outside of time. Perkins had driven himself, and been driven by Chase, beyond reintegration. And yet Perkins was filled with joy! Within himself, Chase Six finally wept. And as he did, his powers fled away in an unending stream like the fog of a harsh night evaporating as the morning sun pierced through and through it.

Keeper Chase's great time-based powers failed as the emergency abated. Unable to maintain the temporal gradient without the urgency of the storm, he could no longer retrieve or even sustain his vast information resource in any practical sense. Yet he had defeated us and dominated the continuum at almost every moment of his adventure. Staggered after his second question and the implications of our answers, he went on to his final and greatest feat. He dared us to prove that he was not an essential evolutionary force within the continuum. Since such a determination would require understanding of other continuums, if such exist, and that necessary understanding would involve an information entropy gradient so vast that it could not even be theoretically sustained, he effectively blocked all further intervention.

But in the end he could not save his youngest crewman. He learned that conscious self-awareness is the most improbable and delicate balance of all within the continuum. Even his great strengths could not bring Surfman Perkins back from the temporal disintegration toward which he had slipped. If the energy-users of Keeper Chase's world group understood how novel and tenuous such consciousness actually is, they would surely behave far differently than they do.

The actual effects of his alterations within the continuum will only become evident in distant times and through much statistical activity on our part. But his greatest effect was the introduction of his third question, to which we may never have a complete or satisfactory answer. Of course, the so-called heroes of Keeper Chase's world group always have this as their primary purpose—that is, the introduction of central and intractable questions.

At a little after nine in the morning of the twenty-first of January, the Coskata boat was sighted through the fading snow from the bluffs of Siasconset, eight miles south of the shore they had left the day before. Soon the entire community was out on the beach, silently watching the surfboat moving toward them, steered by Walter Chase standing at the stern.

The 'Sconset schoolteacher, a young, thin man who had spent two years reading literature at Harvard College, ran up the bluff, a dozen children behind him. As he topped the rise, the thin sun suddenly pierced the damp air and illuminated the tiny boat and its huge captain, looming back even a half mile out.

"Godfreys mighty!" exclaimed the young man to no one in particular. "It's Captain Ahab himself!" for he believed that literature and life were contiguous.

"Naw, tain't," said Widow Tilton. "Hit's Skipper Chase and the Coskata surfboat." She turned to stare at the young man and laughed. "Hit's the only red surfboat around. Skipper painted it red after the Muskeget boat was almost lost in the ice last December, 'cause no one could see it. They wrote from Boston. Said it was nonregulation. Skipper Chase, he wrote back. Don't remember all he wrote, but there was something in his letter about them desk navigators whose experience with ice amounted to sucking it out of their whiskey and sodas at lunch."

The schoolteacher had been only half listening, but now he turned and grinned at Widow Tilton. "He said that to them, did he?" The young man stared again at the approaching boat and then ran down the sand hill. "Come on, boys!" he shouted back at his class. "Let's help get this boat up!"

The Coskata boat grounded silently in a long swell, and a huge crowd waded into the backwash and pulled her up the slick sand. Everyone tried to help the men get out, yet still no one cheered. Instead, soft and kind words flew everywhere, and joy and comfort seemed to warm the very beach.

Walter Chase boomed at the Macy boys to get their oxen and haul the boat up to the dunes. Then he turned and saw Perkins helped and held by Abby Coffin. The boy could no longer speak. Chase smiled at the girl. "Abby, don't take him home. Get him to your sister-in-law's house here in 'Sconset and put him to bed. Get him warm, quick as you can!"

But Abby knew. She could see the emptiness in Roland Perkin's eyes, his fevered cheeks. She wept, so full of grief and pride and love that she could not speak either. But always afterwards she remembered how sharp and yet sad Skipper Chase's eyes had been when he spoke to her and how completely he dominated the beach in those moments at the end of the rescue.

"Isaiah!" shouted Walter Chase. The youngster dashed up, beaming all over his face, so proud that Skipper Chase had picked him out of the great crowd.

"Yessir, Skipper!" He grinned.

"How's that hoss of yours, Isaiah?" said Walter Chase, and now he grinned too.

"Fastest hoss on Nantucket, Skipper," replied the boy promptly. "She'll win at the summer fair for sure!"

"Well, you climb aboard that nag and hustle for town. Find my wife and tell her we got back safe. Then find the rest of them. You know where the crew's folks live?"

The boy nodded and dashed off. Everyone was now moving up the beach toward the village. Each crewman of the *Kirkham* or man of Coskata was surrounded by residents helping them along, throwing coats or blankets over their shoulders, talking at them about the impossible miracle of the rescue.

Captain McCloud of the *Kirkham* staggered along between his huge black crewman and the Widow Tilton, herself well over six feet and two hundred and fifty pounds. Suddenly the old man pitched forward on his knees, pulling the weakened Negro down with him. "Dear God!" he shouted. "Thank Thee for this deliverance! Thank Thee for sparing Thy humble servants. Thank Thee . . ."

Widow Tilton pulled the old man to his feet and, looking back, saw Walter Chase, huge against the dull sun, his tiny eyes like daytime stars. "You better not worry about thanking God, mister," she suddenly said loudly. "It was Skipper Chase got you back here, and don't ever forget that!"

"Walter." It was his uncle beside him. "When they said you was coming in, I put on a gallon of coffee. Come on. Why, man, you're shaking like a leaf!"

Indeed, Walter Chase suddenly was shaking. He could not stop it, and he let his uncle lead him over the dune and down to the little house with its roaring driftwood-filled fire and the huge blackened pot of powerful coffee.

"Uncle," said Walter Chase as he sipped from a huge mug. "I'm shaking so damn much I've got to drink this outside."

He opened the door and stepped back into the narrow, rutted 'Sconset street just as Joe Remsen, sharp in his blue uniform and issue cap, driving the dapper black and gold-trimmed buggy of the Light House Service, pulled by a smart, high-stepping bay, whirled around the corner and pulled up short in a cloud of dust.

"God in Heaven, Walter!" shouted Joe Remsen. "You all did get back!"

Walter Chase, his huge hands still shaking continuously in the thin, cold morning, looked smiling up at his old friend. "Joe, that's just the handsomest one-hoss outfit on the island," he said simply.

"Walter, they say she came apart less than an hour after you got them off! I saw her masts go down at noon yesterday from the tower!"

Walter Chase stretched suddenly and stared, quite piercingly, back at Joe Remsen. "Well," he said, "we didn't need her after the crew got off, did we, Joe?"

At the time his old friend thought Walter Chase was joking, and he laughed out loud. But thinking back on that moment in later years, he realized that Walter Chase had meant what he said. The *Kirkham* had been allowed to collapse because she somehow wasn't *needed* any more. Yet he never asked about it again, but only wondered.

Joe Remsen climbed down from the buggy and shook his head. "We figured you were goners. That damn tug went as far as Great Point and then turned back last night. Too blamed rough, they said, the rotten cowards! By God, Walter, there won't never be another rescue like this one! You better believe that! They're going to build that canal one of these days. Them gasoline engines'll get better and they'll put them in the surf-boats. God Almighty, you took seven of them off. Not one lost. Twenty-six hours out in that smother! It's a miracle! Why, man, you moved heaven and earth . . ."

The hot coffee drained its warmth through Walter Chase, and suddenly he felt drowsy. "Joe, we never did try a drail for squeteague out there. Just too blamed busy the whole time . . ."

And the two old friends grinned and chuckled at each other in the winter sunlight on a 'Sconset street.

Resolution of the Paradox: A Philosophical Puppet Play

Abner Shimony

Dramatis personae: Zeno, Pupil, Lion

Scene: The school of Zeno at Elea.

Pup. Master! There is a lion in the streets!

Zen. Very good. You have learned your lesson in geography well. The fifteenth meridian, as measured from Greenwich, coincides with the high road from the Temple of Poseidon to the Agora—but you must not forget that it is an imaginary line.

Pup. Oh no, Master! I must humbly disagree. It is a *real* lion, a *menagerie* lion, and it is coming toward the school!

Zen. My boy, in spite of your proficiency at geography, which is commendable in its way—albeit essentially the art of the surveyor and hence separated by the hair of the theodolite from the craft of a slave—you are deficient in philosophy. That which is real cannot be imaginary, and that which is imaginary cannot be real. Being is, and non-being is not, as my revered teacher Parmenides demonstrated first, last, and continually, and as I have attempted to convey to you.

Pup. Forgive me, Master. In my haste and excitement, themselves expressions of passion unworthy of you and of our school, I have spoken obscurely. Into the gulf between the thought and the word, which, as you have taught us, is the trap set by non-being, I have again fallen. What I meant to say is that a lion has escaped from the zoo, and with deliberate speed it is rushing in the direction of the school and soon will be here!

The lion appears in the distance.

Zen. O my boy, my boy! It pains me to contemplate the impenetrability of the human intellect and its incommensurability with the truth. Furthermore, I now recognize that a thirty-year novitiate is too brief—*sub specie aeternitatis*—and must be extended to forty years, before the apprenticeship proper can begin. A real lion, perhaps; but really running, impossible; and really arriving here, absurd!

Pup. Master . . .

Zen. In order to run from the zoological garden to the Eleatic school, the lion would first have to traverse half the distance.

The lion traverses half the distance.

Zen. But there is a first half of that half, and a first half of that half, and yet again a first half of that half to be traversed. And so the halves would of necessity regress to the first syllable of recorded time—nay, they would recede yet earlier than the first syllable. To have traveled but a minute part of the interval from the zoological garden to the school, the lion would have been obliged to embark upon his travels *infinitely long ago.*

The lion bursts into the schoolyard.

Pup. O Master, run, run! He is upon us!

Zen. And thus, by *reductio ad absurdum,* we have proved that the lion could never have *begun* the course, the mere fantasy of which has so unworthily filled you with panic.

The pupil climbs an Ionic column, while the lion devours Zeno.

Pup. My mind is in a daze. Could there be a flaw in the Master's argument?

Parallelism

Piet Hein

To Martin Gardner

"Lines that are parallel
meet at Infinity!"
Euclid repeatedly,
heatedly,
  urged
Until he died.
and so reached that vicinity:
in it he
found that the damned things
  diverged.

Prelude . . .

Douglas Hofstadter

(from *Gödel, Escher, Bach: An Eternal Golden Braid*)

(Achilles and the Tortoise have come to the residence of their friend the Crab, to make the acquaintance of one of his friends, the Anteater. The introductions having been made, the four of them settle down to tea.)

Tortoise: We have brought along a little something for you, Mr. Crab.

Crab: That's most kind of you. But you shouldn't have.

Tortoise: Just a token of our esteem. Achilles, would you like to give it to Mr. C?

Achilles: Surely. Best wishes, Mr. Crab. I hope you enjoy it.

(Achilles hands the Crab an elegantly wrapped present, square and very thin. The Crab begins unwrapping it.)

Anteater: I wonder what it could be.

Crab: We'll soon find out. *(Completes the unwrapping, and pulls out the gift.)* Two records! How exciting! But there's no label. Uh-oh—is this another of your "specials," Mr. T?

Tortoise: If you mean a phonograph-breaker, not this time. But it is in fact a custom-recorded item, the only one of its kind in the entire world. In fact, it's never even been heard before—except, of course, when Bach played it.

Crab: When Bach played it? What do you mean, exactly?

Achilles: Oh, you are going to be fabulously excited, Mr. Crab, when Mr. T tells you what these records in fact are.

Tortoise: Oh, you go ahead and tell him, Achilles.

Achilles: May I? Oh, boy! I'd better consult my notes, then. *(Pulls out a small filing card, and clears his voice.)* Ahem. Would you be interested in hearing about the remarkable new result in mathematics, to which your records owe their existence?

Crab: My records derive from some piece of mathematics? How curious! Well, now that you've provoked my interest, I must hear about it.

Achilles: Very well, then. *(Pauses for a moment to sip his tea, then resumes.)* Have you heard of Fermat's infamous "Last Theorem"?

Anteater: I'm not sure . . . It sounds strangely familiar, and yet I can't quite place it.

Achilles: It's a very simple idea. Pierre de Fermat, a lawyer by vocation but mathematician by avocation, had been reading in his copy of the classic text *Arithmetica* by Diophantus, and came across a page containing the equation

$$a^2 + b^2 = c^2$$

He immediately realized that this equation has infinitely many solutions a, b, c, and then wrote in the margin the following notorious comment:

> The equation
>
> $$a^n + b^n = c^n$$
>
> has solutions in positive integers a, b, c, and n only when $n = 2$ (and then there are infinitely many triplets a, b, c which satisfy the equation): but there are no solutions for $n > 2$. I have discovered a truly marvelous proof of this statement, which, unfortunately, this margin is too small to contain.

Even since that day, some three hundred years ago, mathematicians have been vainly trying to do one of two things: either to prove Fermat's claim, and thereby vindicate Fermat's reputation, which, although very high, has been somewhat tarnished by skeptics who think he never really found the proof he claimed to have found—or else to refute the claim, by finding a counterexample: a set of four integers a, b, c, and n, with $n > 2$, which satisfy the equation. Until very recently, every attempt in either direction had met with failure. To be sure, the Theorem has been proven for many specific values of n—in particular, all n up to 125,000.

Anteater: Shouldn't it be called a "Conjecture" rather than a "Theorem," if it's never been given a proper proof?

Achilles: Strictly speaking, you're right, but tradition has kept it this way.

Crab: Has someone at last managed to resolve this celebrated question?

Achilles: Indeed! In fact, Mr. Tortoise has done so, and as usual, by a wizardly stroke. He has not only found a PROOF of Fermat's Last Theorem (thus justifying its name as well as vindicating Fermat), but also a COUNTEREXAMPLE, thus showing that the skeptics had good intuition!

Crab: Oh my gracious! That is a revolutionary discovery.

Anteater: But please don't leave us in suspense. What magical integers are they, that satisfy Fermat's equation? I'm especially curious about the value of n.

Achilles: Oh, horrors! I'm most embarrassed! Can you believe this? I left the values at home on a truly colossal piece of paper. Unfortunately it was too huge to bring along. I wish I had them here to show to you. If it's of any help to you, I do remember one thing—the value of n is the only positive integer which does not occur anywhere in the continued fraction for π.

Crab: Oh, what a shame that you don't have them here. But there's no reason to doubt what you have told us.

Anteater: Anyway, who needs to see n written out decimally? Achilles has just told us how to find it. Well, Mr. T, please accept my hearty felicitations, on the occasion of your epoch-making discovery!

Tortoise: Thank you. But what I feel is more important than the result itself is the practical use to which my result immediately led.

Crab: I am dying to hear about it, since I always thought number theory was the Queen of Mathematics—the purest branch of mathematics—the one branch of mathematics which has NO applications!

Tortoise: You're not the only one with that belief, but in fact it is quite impossible to make a blanket statement about when or how some branch—or even some individual Theorem—of pure mathematics will have important repercussions outside of mathematics. It is quite unpredictable—and this case is a perfect example of that phenomenon.

Achilles: Mr. Tortoise's double-barreled result has created a breakthrough in the field of acoustico-retrieval!

Anteater: What is acoustico-retrieval?

Achilles: The name tells it all: it is the retrieval of acoustic information from extremely complex sources. A typical task of acoustico-retrieval is to reconstruct the sound which a rock made on plummeting into a lake from the ripples which spread out over the lake's surface.

Crab: Why, that sounds next to impossible!

Achilles: Not so. It is actually quite similar to what one's brain does, when it reconstructs the sound made in the vocal cords of another person from the vibrations transmitted by the eardrum to the fibers in the cochlea.

Crab: I see. But I still don't see where number theory enters the picture, or what this all has to do with my new records.

Achilles: Well, in the mathematics of acoustico-retrieval, there arise many questions which have to do with the number of solutions of certain Diophantine equations. Now Mr. T has been for years trying to find a way of reconstructing the sounds of Bach playing his harpsichord, which took place over two hundred years ago, from calculations involving the motions of all the molecules in the atmosphere at the present time.

Anteater: Surely that is impossible! They are irretrievably gone, gone forever!

Achilles: Thus think the naïve . . . But Mr. T has devoted many years to this problem, and came to the realization that the whole thing hinged on the number of solutions to the equation

$$a^n + b^n = c^n$$

in positive integers, with $n > 2$.

Tortoise: I could explain, of course, just how this equation arises, but I'm sure it would bore you.

Achilles: It turned out that acoustico-retrieval theory predicts that the Bach sounds can be retrieved from the motion of all the molecules in the atmosphere, provided that EITHER there exists at least one solution to the equation—

Crab: Amazing!

Anteater: Fantastic!

Tortoise: Who would have thought!

Achilles: I was about to say, "provided that there exists EITHER such a solution OR a proof that there are NO solutions!" And therefore, Mr. T, in careful fashion, set about working at both ends of the problem, simultaneously. As it turns out, the discovery of the counterexample was the key ingredient to finding the proof, so the one led directly to the other.

Crab: How could that be?

Tortoise: Well, you see, I had shown that the structural layout of any proof of Fermat's Last Theorem—if one existed—could be described by an elegant formula, which, it so happened, depended on the values of a solution to a certain equation. When I found this second equation, to my surprise it turned out to be the Fermat equation. An amusing accidental relationship between form and content. So when I found the counterexample, all I needed to do was to use those numbers as a blueprint for constructing my proof that there were no solutions to the equation. Remarkably simple, when you think about it. I can't imagine why no one had ever found the result before.

Achilles: As a result of this unanticipatedly rich mathematical success, Mr. T was able to carry out the acoustico-retrieval which he had so long dreamed of. And Mr. Crab's present here represents a palpable realization of all this abstract work.

Crab: Don't tell me it's a recording of Bach playing his own works for harpsichord!

Achilles: I'm sorry, but I have to, for that is indeed just what it is! This is a set of two records of Johann Sebastian Bach playing all of his *Well-Tempered Clavier.* Each record contains one of the two volumes of the *Well-Tempered Clavier;* that is to say, each record contains 24 preludes and fugues—one in each major and minor key.

Crab: Well, we must absolutely put one of these priceless records on, immediately! And how can I ever thank the two of you?

Tortoise: You have already thanked us plentifully, with this delicious tea which you have prepared.

(The Crab slides one of the records out of its jacket, and puts it on. The sound of an incredibly masterful harpsichordist fills the room, in the highest imaginable fidelity. One even hears—or is it one's imagination?—the soft sounds of Bach singing to himself as he plays . . .)

Crab: Would any of you like to follow along in the score? I happen to have a unique edition of the *Well-Tempered Clavier,* specially illuminated by a teacher of mine who happens also to be an unusually fine calligrapher.

Tortoise: I would very much enjoy that.

(The Crab goes to his elegant glass-enclosed wooden bookcase, opens the doors, and draws out two large volumes.)

Crab: Here you are, Mr. Tortoise. I've never really gotten to know all the beautiful illustrations in this edition. Perhaps your gift will provide the needed impetus for me to do so.

Tortoise: I do hope so.

Anteater: Have you ever noticed how in these pieces the prelude always sets the mood perfectly for the following fugue?

Crab: Yes. Although it may be hard to put it into words, there is always some subtle relation between the two. Even if the prelude and fugue do not have a common melodic subject, there is nevertheless always some intangible abstract quality which underlies both of them, binding them together very strongly.

Tortoise: And there is something very dramatic about the few moments of silent suspense hanging between prelude and fugue—that moment where the the theme of the fugue is about to ring out, in single tones, and then to join with itself in ever-increasingly complex levels of weird, exquisite harmony.

Achilles: I know just what you mean. There are so many preludes and fugues which I haven't yet gotten to know, and for me that fleeting interlude of silence is very exciting; it's a time when I try to second-guess old Bach. For example, I always wonder what the fugue's tempo will be: allegro, or adagio? Will it be in 6/8, or 4/4? Will it have three voices, or five—or four? And then, the first voice starts . . . Such an exquisite moment.

Crab: Ah, yes, well do I remember those long-gone days of my youth, the days when I thrilled to each new prelude and fugue, filled with the excitement of their novelty and beauty and the many unexpected surprises which they conceal.

Achilles: And now? Is that thrill all gone?

Crab: It's been supplanted by familiarity, as thrills always will be. But in that familiarity there is also a kind of depth, which has its own compensations. For instance, I find that there are always new surprises which I hadn't noticed before.

Achilles: Occurrences of the theme which you had overlooked?

Crab: Perhaps—especially when it is inverted and hidden among several other voices, or where it seems to come rushing up from the depths, out of nowhere. But there are also amazing modulations which it is marvelous to listen to over and over again, and wonder how old Bach dreamt them up.

Achilles: I am very glad to hear that there is something to look forward to, after I have been through the first flush of infatuation with the *Well-Tempered Clavier*—although it also makes me sad that this stage could not last forever and ever.

Crab: Oh, you needn't fear that your infatuation will totally die. One of the nice things about that sort of youthful thrill is that it can always be resuscitated, just when you thought it was finally dead. It just takes the right kind of triggering from the outside.

Achilles: Oh, really? Such as what?

Crab: Such as hearing it through the ears, so to speak, of someone to whom it is a totally new experience—someone such as you, Achilles. Somehow the excitement transmits itself, and I can feel thrilled again.

Achilles: That is intriguing. The thrill has remained dormant somewhere inside you, but by yourself, you aren't able to fish it up out of your subconscious.

Crab: Exactly. The potential of reliving the thrill is "coded," in some unknown way, in the structure of my brain, but I don't have the power to summon it up at will; I have to wait for chance circumstance to trigger it.

Achilles: I have a question about fugues which I feel a little embarrassed about asking, but as I am just a novice at fugue-listening, I was wondering if perhaps one of you seasoned fugue-listeners might help me in learning . . . ?

Tortoise: I'd certainly like to offer my own meager knowledge, if it might prove of some assistance.

Achilles: Oh, thank you. Let me come at the question from an angle. Are you familiar with the print called *Cube with Magic Ribbons,* by M. C. Escher?

Tortoise: In which there are circular bands having bubble-like distortions which, as soon as you've decided that they are bumps, seem to turn into dents—and vice versa?

Achilles: Exactly.

Crab: I remember that picture. Those little bubbles always seem to flip back and forth between being concave and convex, depending on the direction that you approach them from. There's no way to see them simultaneously as concave AND convex—somehow one's brain doesn't allow that. There are two mutually exclusive "modes" in which one can perceive the bubbles.

Achilles: Just so. Well, I seem to have discovered two somewhat analogous modes in which I can listen to a fugue. The modes are these: either to follow one individual voice at a time, or to listen to the total effect of all of them together, without trying to disentangle one from another. I have tried out both of these modes, and, much to my frustration, each one of them shuts out the other. It's simply not in my power to follow the paths of individual voices and at the same time to hear the whole effect. I find that I flip back and forth between one mode and the other, more or less spontaneously and involuntarily.

Anteater: Just as when you look at the magic bands, eh?

Achilles: Yes, I was just wondering . . . does my description of these two modes of fugue-listening brand me unmistakably as a naïve, inexperienced listener, who couldn't even begin to grasp the deeper modes of perception which exist beyond his ken?

Tortoise: No, not at all, Achilles. I can only speak for myself, but I too find myself shifting back and forth from one mode to the other without exerting any conscious control over which mode should be dominant. I don't know if our other companions here have also experienced anything similar.

Crab: Most definitely. It's quite a tantalizing phenomenon, since you feel that the essence of the fugue is flitting about you, and you can't quite grasp all of it, because you can't quite make yourself function both ways at once.

Anteater: Fugues have that interesting property, that each of their voices is a piece of music in itself; and thus a fugue might be thought of as a collection of several distinct pieces of music, all based on one single theme, and all played simultaneously. And it is up to the listener (or his subconscious) to decide whether it should be perceived as a unit, or as a collection of independent parts, all of which harmonize.

Achilles: You say that the parts are "independent," yet that can't be literally true. There has to be some coordination between them, otherwise when they were put together one would just have an unsystematic clashing of tones—and that is as far from the truth as could be.

Anteater: A better way to state it might be this: if you listened to each voice on its own, you would find that it seemed to make sense all by itself. It could stand alone, and that is the sense in which I meant that it is independent. But you are quite right in pointing out that each of these individually meaningful lines fuses with the others in a highly nonrandom way, to make a graceful totality. The art of writing a beautiful fugue lies precisely in this ability, to manufacture several different lines, each one of which gives the illusion of having been written for its own beauty, and yet which when taken together form a whole, which does not feel forced in any way. Now, this dichotomy between hearing a fugue as a whole, and hearing its component voices, is a particular example of a very general dichotomy, which applies to many kinds of structures built up from lower levels.

Achilles: Oh, really? You mean that my two "modes" may have some more general type of applicability, in situations other than fugue-listening?

Anteater: Absolutely.

Achilles: I wonder how that could be. I guess it has to do with alternating between perceiving something as a whole, and perceiving it as a collection of parts. But the only place I have ever run into that dichotomy is in listening to fugues.

Tortoise: Oh, my, look at this! I just turned the page while following the music, and came across this magnificent illustration facing the first page of the fugue.

Crab: I have never seen that illustration before. Why don't you pass it 'round?

(The Tortoise passes the book around. Each of the foursome looks at it in a characteristic way—this one from afar, that one from close up, everyone tipping his head this way and that in puzzlement. Finally it has made the rounds, and returns to the Tortoise, who peers at it rather intently.)

Achilles: Well, I guess the prelude is just about over. I wonder if, as I listen to this fugue, I will gain any more insight into the question, "What is the right way to listen to a fugue: as a whole, or as the sum of its parts?"

Tortoise: Listen carefully, and you will!

(The prelude ends. There is a moment of silence; and . . .

[ATTACCA]

The Third Sally, or The Dragons of Probability

Stanislaw Lem

(from *The Seven Sallies of Trurl and Klapaucius*)

Trurl and Klapaucius were former pupils of the great Cerebron of Umptor, who for forty-seven years in the School of Higher Neantical Nillity expounded the General Theory of Dragons. Everyone knows that dragons don't exist. But while this simplistic formulation may satisfy the layman, it does not suffice for the scientific mind. The School of Higher Neantical Nillity is in fact wholly unconcerned with what *does* exist. Indeed, the banality of existence has been so amply demonstrated, there is no need for us to discuss it any further here. The brilliant Cerebron, attacking the problem analytically, discovered three distinct kinds of dragon: the mythical, the chimerical, and the purely hypothetical. They were all, one might say, nonexistent, but each nonexisted in an entirely different way. And then there were the imaginary dragons, and the a-, anti- and minus-dragons (colloquially termed nots, noughts and oughtn'ts by the experts), the minuses being the most interesting on account of the well-known dracological paradox: when two minuses hypercontiguate (an operation in the algebra of dragons corresponding roughly to simple multiplication), the product is 0.6 dragon, a real nonplusser. Bitter controversy raged among the experts on the question of whether, as half of them claimed, this fractional beast began from the head down or, as the other half maintained, from the tail up. Trurl and Klapaucius made a great contribution by showing the error of both positions. They were the first to apply probability theory to this area and, in so doing, created the field of statistical draconics, which says that dragons are thermodynamically impossible only in the probabilistic sense, as are elves, fairies, gnomes, witches, pixies and the like. Using the general equation of improbability, the two constructors obtained the coefficients of pixation, elfinity, kobolding, etc. They found that for the spontaneous manifestation of an average dragon, one would have to wait a good sixteen quintoquadrillion heptillion years. In other words, the whole problem would have remained a mathematical curiosity had it not been for that famous tinkering passion of Trurl, who decided to examine the nonphenomenon empirically. First, as he was dealing with the highly improbable, he invented a probability amplifier and ran tests in his basement—then later at the Dracogenic Proving Grounds established and funded by the Academy. To this day those who (sadly enough) have no knowledge of the General Theory of Improbability ask why Trurl probabilized a dragon and not an elf or goblin. The answer is simply that dragons are more probable than elves or goblins to begin with. True, Trurl might have gone further with his amplifying experiments, had not the first been so discouraging—discouraging in that the materialized dragon tried to make a meal of him. Fortunately, Klapaucius was nearby and lowered the probability, and the monster vanished. A number of scholars subsequently repeated the experiment on a phantasmatron, but, as they lacked the necessary know-how and sang-froid, a considerable quantity of dragon spawn, raising an ungodly perturbation, broke loose. Only then did it become clear that those odious beasts enjoyed an existence quite different from that of ordinary cupboards, tables and chairs; for dragons are distinguished by their probability rather than by their actuality, though granted, that probability is overwhelming once they've actually come into being. Suppose, for example, one organizes a hunt for such a dragon, surrounds it, closes in, beating the brush. The circle of sportsmen, their weapons cocked and ready, finds only a burnt patch of earth and an unmistakable smell: the dragon, seeing itself cornered, has slipped from real to configurational space. An extremely obtuse and brutal creature, it does this instinctively, of course. Now, ignorant and backward persons will occasionally demand that you show them this configurational space of yours, apparently unaware that electrons, whose existence no one in his right mind would question, also move exclusively in configurational space, their comings and goings fully dependent on curves of probability. Though it is easier not to believe in electrons than in dragons: electrons, at least taken singly, won't try to make a meal of you.

A colleague of Trurl, one Harborizian Cybr, was the first to quantize a dragon, detecting a particle known as the dracotron, the energy of which is measured—obviously—in units of dracon by a dracometer, and he even determined the coordinates of its tail, for which he nearly paid with his life. Yet what did these scientific achievements concern the common folk, who were now greatly harassed by dragons ranging the countryside, filling the air with their howls and flames and trampling, and in places even exacting tribute in the form of young virgins? What did it concern the poor villagers that Trurl's dragons, indeterministic hence heuristic, were behaving exactly according to theory though contrary to all notions of decency, or that his theory could predict the curve of the tails that demolished their barns and leveled their crops? It is not surprising, then, that the general public, instead of appreciating the value of Trurl's revolutionary invention, held it much against him. A group of individuals thoroughly benighted in matters of science waylaid the famous constructor and gave him a good thrashing. Not this that deterred him and his friend Klapaucius from further experimentation, which showed that the extent of a dragon's existence depends mainly on its whim, though also on its degree of satiety, and that the only sure method of negating it is to reduce the probability to zero or lower. All this research, naturally enough, took a great deal of time and energy; meanwhile the dragons that had gotten loose were running rampant, laying waste to a variety of planets and moons. What was worse, they multiplied. Which enabled Klapaucius to publish an excellent article entitled "Covariant Transformation from Dragons to Dragonets, in the Special Case of Passage from States Forbidden by the Laws of Physics to Those

Forbidden by the Local Authorities." The article created a sensation in the scientific world, where there was still talk of the amazing polypolice beast that had been used by the intrepid constructors against King Krool to avenge the deaths of their colleagues. But far greater was the sensation caused by the news that a certain constructor known as Basiliscus the Gorgonite, traveling through the Galaxy, was apparently making dragons appear by his presence—and in places where no one had ever seen a dragon before. Whenever the situation grew desperate and catastrophe seemed imminent, this Basiliscus would turn up, approach the sovereign of that particular area and, settling on some outrageous fee after long hours of bargaining, would undertake to extirpate the beasts. At which he usually succeeded, though no one knew quite how, since he worked in secret and alone. True, the guarantee he offered for dragon removal—dracolysis—was only statistical; though one ruler did pay him in similar coin, that is, in ducats that were only statistically good. After that, the insolent Basiliscus always used aqua regia to check the metallic reliability of his royal payments. One sunny afternoon Trurl and Klapaucius met and held the following conversation:

"Have you heard about this Basiliscus?" asked Trurl.

"Yes."

"Well, what do you think?"

"I don't like it."

"Nor do I. How do you suppose he does it?"

"With an amplifier."

"A probability amplifier?"

"Either that, or oscillating fields."

"Or a paramagnedracic generator."

"You mean, a draculator?"

"Yes."

"Ah."

"But really," cried Trurl, "that would be criminal! That would mean he was bringing the dragons with him, only in a potential state, their probability near zero; then, after landing and getting the lay of the land, he was increasing the chances, raising the potential, strengthening the probability until it was almost a certainty. And then, of course, you have virtualization, materialization, full manifestation."

"Of course. And he probably shuffles the letters of the matrix to make the dragons *grand*."

"Yes, and the poor people *groan* in *agony* and *gore*. Terrible!"

"What do you think; does he then apply an irreversible antidraconian retroectoplasmatron, or simply lower the probability and walk off with the gold?"

"Hard to say. Though if he's only improbabilizing, that would be an even greater piece of villainy, since sooner or later the fractional fluctuations would have to give rise to a draconic iso-oscillation—and the whole thing would start all over again."

"Though by that time both he and the money would be gone," observed Klapaucius.

"Shouldn't we report him to the Main Office?"

"Not just yet. He may not be doing this, after all. We have no real proof. Statistical fluctuations can occur without an amplifier; at one time, you know, there were neither amplifiers nor phantasmatrons, yet dragons did appear. Purely on a random basis."

"True . . ." replied Trurl. "But *these* appear immediately after he arrives on the planet!"

"I know. Still, reporting a fellow constructor—it just isn't done. Though there's no reason we can't take measures of our own."

"No reason at all."

"I'm glad you agree. But what exactly should we do?"

At this point the two famous dracologists got into a discussion so technical, that anyone listening in wouldn't have been able to make head or tail of it. There were such mysterious words as "discontinuous orthodragonality," "grand draconical ensembles," "high-frequency binomial fafneration," "abnormal saurian distribution," "discrete dragons," "indiscrete dragons," "drasticodracostochastic control," "simple Grendelian dominance," "weak interaction dragon diffraction," "aberrational reluctance," "informational figmentation," and so on.

The upshot of all this penetrating analysis was the third sally, for which the constructors prepared most carefully, not failing to load their ship with a quantity of highly complicated devices.

In particular they took along a scatter-scrambler and a special gun that fired negative heads. After landing on Eenica, then on Meenica, then finally on Mynamoaca, they realized it would be impossible to comb the whole infested area in this way and they would have to split up. This was most easily done, obviously, by separating; so after a brief council of war each set out on his own. Klapaucius worked for a spell on Prestopondora for the Emperor Maximillion, who was prepared to offer him his daughter's hand in marriage if only he would get rid of those vile beasts. Dragons of the highest probability were everywhere, even in the streets of the capital, and the place literally swarmed with virtuals. A virtual dragon, the uneducated and simple-minded might say, "isn't really there," having no observable substance nor displaying the least intention of acquiring any; but the Cybr-Trurl-Klapaucius-Leech calculation (not to mention the Drachendranginger wave equation) clearly shows that a dragon can jump from configurational to real space with no more effort than it takes to jump off a cliff. Thus, in any room, cellar or attic, provided the probability is high, you could meet with a dragon or possibly even a metadragon.

Instead of chasing after the beasts, which would have accomplished little or nothing, Klapaucius, a true theoretician, approached the problem methodically; in squares and promenades, in barns and hostels he placed probabilistic battery-run dragon dampers, and in no time at all the beasts were extremely rare. Collecting his fee, plus an honorary degree and an engraved loving cup, Klapaucius blasted off to rejoin his friend. On the way, he noticed a planet and someone waving to him frantically. Thinking it might be Trurl in some sort of trouble, he landed. But it was only the inhabitants of Truf-flandria, the subjects of King Pfftius, gesticulating. The Truf-flandrians held to various superstitions and primitive beliefs; their religion, Pneumatological Dracolatry, taught that dragons appeared as a divine retribution for their sins and took possession of all unclean souls. Quickly realizing it would be useless to enter into a discussion with the royal dracologians—their methods consisted primarily of waving censers and distributing sacred relics—Klapaucius instead conducted soundings of the outlying terrain. These revealed the planet was occupied by only one beast, but that beast belonged to the terrible genus of Echidnosaurian hypervipers. He offered the King his services. The King, however, answered in a vague, roundabout fashion, evidently under the influence of that ridiculous doctrine which would have the origin of dragons be somehow supernatural. Perusing the local newspapers, Klapaucius learned that the dragon terrorizing the planet was considered by some to be a single thing, and by others, a multiplex creature that could operate in several locations at the same time. This gave him pause—though it wasn't so surprising really, when you considered that the localization of these odious phenomena was subject to so-called dragonomalies, in which certain specimens, particularly when abstracted, underwent a "smearing" effect, which was in reality nothing more than a simple isotopic spin acceleration of asynchronous quantum moments. Much as a hand, emerging from the water fingers-first, appears above the surface in the form of five seemingly separate and independent items, so do dragons, emerging from the lairs of their configurational space, on occasion appear to be plural, though in point of fact they are quite singular. Towards the end of his second audience with the King, Klapaucius inquired if perhaps Trurl were on the planet and gave a detailed description of his comrade. He was astonished to hear that yes, his comrade had only recently visited their kingdom and had even undertaken to exorcise the monster, had in fact accepted a retainer and departed for the neighboring mountains where the monster had been most frequently sighted. Had then returned the next day, demanding the rest of his fee and presenting four and twenty dragon's teeth as proof of his success. There was some misunderstanding, however, and it was decided to withhold payment until the matter was fully cleared up. At which Trurl flew into a rage and in a loud voice made certain comments about His Royal Highness that were perilously close to lèse majesty if not treason, then stormed out without leaving a forwarding address. That very same day the monster reappeared as if nothing had happened and, alas, ravaged their farms and villages more fiercely than before.

Now this story seemed questionable to Klapaucius, though on the other hand it was hard to believe the good King was lying, so he packed his knapsack with all sorts of powerful dragon-exterminating equipment and set off for the mountains, whose snowcapped peaks rose majestically in the east.

It wasn't long before he saw dragon prints and got an unmistakable whiff of brimstone. On he went, undaunted, holding his weapon in readiness and keeping a constant eye on the needle of his dragon counter. It stayed at zero for a spell, then began to give nervous little twitches, until, as if struggling with itself, it slowly crawled towards the number one. There was no doubt now: the Echidnosaur was close at hand. Which amazed Klapaucius, for he couldn't understand how his trusty friend and renowned theoretician, Trurl, could have gotten so fouled up in his calculations as to fail to wipe the dragon out for good. Nor could he imagine Trurl returning to the royal palace and demanding payment for what he had not accomplished.

Klapaucius then came upon a group of natives. They were plainly terrified, the way they kept looking around and trying to stay together. Bent beneath heavy burdens balanced on their backs and heads, they were stepping single-file up the mountainside. Klapaucius accosted the procession and asked the first native what they were about.

"Sire!" replied the native, a lower court official in a tattered tog and cummerbund. " 'Tis the tribute we carry to the dragon."

"Tribute? Ah yes, the tribute! And what *is* the tribute?"

"Nothin' more 'r less, Sire, than what the dragon would have us bring it: gold coins, precious stones, imported perfumes, an' a passel o' other valuables."

This was truly incredible, for dragons never required such tributes, certainly not perfume—no perfume could ever mask their own natural fetor—and certainly not currency, which was useless to them.

"And does it ask for young virgins, my good man?" asked Klapaucius.

"Virgins? Nay, Sire, tho' there war a time . . . we had to cart 'em in by the bevy, we did. . . . Only that war before the stranger came, the furrin gentleman, Sire, a-walkin' around the rocks with 'is boxes an' contraptions, all by 'isself. . . ." Here the worthy native broke off and stared at the instruments and weapons Klapaucius was carrying, particularly the large dragon counter that was ticking softly all the while, its red pointer jumping back and forth across the white dial.

"Why, if he dinna have one . . . jus' like yer Lordship's," he said in a hushed voice. "Aye, jus' like . . . the same wee stiggermajigger and a' the rest . . ."

"There was a sale on them," said Klapaucius, to allay the native's suspicions. "But tell me, good people, do you happen to know what became of this stranger?"

"What became o' him, ye ask? That we know not, Sire, to be sure. 'Twas, if I not mistake me, but a fortnight past— 'twas 'twas not, Master Gyles, a fortnight withal an' nae more?"

" 'Twas, 'twas, 'tis the truth ye speak, the truth aye, a fortnight sure, or maybe two."

"Aye! So he comes to us, yer Grace, partakes of our 'umble fare, polite as ye please an' I'll not gainsay it, nay, a parfit gentleman true, pays hondsomely, inquires after the missus don't y'know, aye an' then he sits 'isself down, spreads out a' them contraptions an' thin's with clocks in 'em, y'see, an scribbles furious-like, numbers they are, one after t'other, in this wee book he keep in 'is breast pocket, then takes out a—whad'yacallit—therbobbiter thingamabob . . ."

"Thermometer?"

"Aye, that's it. A Thermometer . . . an' he says it be for dragons, an' pokes it here an' there, Sire, an' scribbles in 'is book again, then he takes a' them contraptions an' things an' packs 'em up an' puts 'em on 'is back an' says farewell an' goes 'is merry way. We never saw 'im more, yer Honor. That very night we hear a thunder an' a clatter, oh, a good ways off, 'bout as far as Mount Murdigras—'tis the one, Sire, hard by yon peak, aye, that one thar, looks like a hawk, she do, we call 'er Pfftius Peak after our beluved King, an' that one that on t'uther side, bent over like t'would spread 'er arse, that be the Dollymog, which, accordin' to legend—"

"Enough of the mountains, worthy native," said Klaupaucius. "You were saying there was thunder in the night. What happened then?"

"Then, Sire? Why nothin' to be sure. The hut she give a jump an' I falls outta bed, to which I'm well accustomed, mind ye, seein' as how the wicked beast allus comes a-bumpin' gainst the house with 'er tail an' send a feller flyin'—like when Master Gyles' ayn brother londed in the privy 'cause the creatur' gets a hankerin' to scratch 'isself on the corner o' the roof . . ."

"To the point, man, get to the point!" cried Klapaucius. "There was thunder, you fell down, and then what?"

"Then nothin', like I says before an' thought I made it clear. Nothin', an' if'n there war somethin', there'd be somethin', only there war nothin' sure an' that be the long an' the short of it! D'ye agree. Master Gyles?"

"Aye, sure 'tis the truth ye speak, 'tis."

Klapaucius bowed and stepped back, and the whole procession continued up the mountain, the natives straining beneath the dragon's tribute. He supposed they would place it in some cave designated by the beast, but didn't care to ask for details; his head was already spinning from listening to the local official and his Master Gyles. And anyway, he had heard one of the natives say to another that the dragon had chosen "a spot as near us an' as near 'isself as could be found."

Klapaucius hurried on, picked his way according to the readings of the dragonometer he kept on a chain around his neck. As for the counter, its pointer had come to rest on exactly eight-tenths of a dragon.

"What in the devil is it, an indeterminant dragon?" he thought as he marched, stopping to rest every now and then, for the sun beat fiercely and the air was so hot that everything shimmered. There was no vegetation anywhere, not a scrap, only baked mud, rocks and boulders as far as the eye could see.

An hour passed, the sun hung lower in the heavens, and Klapaucius still walked through fields of gravel and scree, through craggy passes, till he found himself in a place of narrow canyons and ravines full of chill and darkness. The red pointer crept to nine-tenths, gave a shudder, and froze.

Klapaucius put his knapsack on a rock and had just taken off his antidragon belt when the indicator began to go wild, so he grabbed his probability extinguisher and looked all around. Situated on a high bluff, he was able to see into the gorge below, where something moved.

"That must be her!" he thought, since Echidnosaurs are invariably female.

Could *that* be why it didn't demand young virgins? But no, the native said it had before. Odd, most odd. But the main thing now, Klapaucius told himself, was to shoot straight and everything would be all right. Just in case, however, he reached for his knapsack again and pulled out a can of dragon repellent and an atomizer. Then he peered over the edge of the rock. At the bottom of the gorge, along the bed of a dried-up stream walked a grayish brown dragoness of enormous proportions, though with sunken sides as if it had been starved. All sorts of thoughts ran through Klapaucius' head. Annihilate the thing by reversing the sign of its pentapendragonal coefficient from positive to negative, thereby raising the statistical probability of its nonexistence over that of its existence? Ah, but how very risky that was, when the least deviation could prove disastrous: more than one poor soul, seeking to produce the lack of a dragon, had ended up instead with the back of the dragon—resulting in a beast with two backs—and nearly died of embarrassment! Besides, total deprobabilization would rule out the possibility of studying the Echidnosaur's behavior. Klapaucious wavered; he could see a splendid dragonskin tacked on the wall of his den, right above the fireplace. But this wasn't the time to indulge in daydreams—though a dracozoologist would certainly be delighted to receive an animal with such unusual tastes. Finally, as Klapaucius got into position, it occurred to him what a nice little article might be written up on the strength of a well-preserved specimen, so he put down the extinguisher, lifted the gun that fired negative heads, took careful aim and pulled the trigger.

The roar was deafening. A cloud of white smoke engulfed Klapaucius and he lost sight of the beast for a moment. Then the smoke cleared.

There are a great many old wives' tales about dragons. It is said, for example, that dragons can sometimes have seven heads. This is sheer nonsense. A dragon can have only one head, for the simple reason that having two leads to disagreements and violent quarrels; the polyhydroids, as the scholars call them, died out as a result of internal feuds. Stubborn and headstrong by nature, dragons cannot tolerate opposition, therefore two heads in one body will always bring about a swift death: each head, purely to spite the other, refuses to eat, then maliciously holds its breath—with the usual consequences. It was this phenomenon which Euphorius Cloy exploited when he invented the anticapita cannon. A small auxiliary electron head is discharged into the dragon's body. This immediately gives rise to irreconcilable differences of opinion and the dragon is immobilized by the ensuing deadlock. Often it will stand there, stiff as a board, for a day, a week, even a month; sometimes a year goes by before the beast will collapse, exhausted. Then you can do with it what you will.

But the dragon Klapaucius shot reacted strangely, to say the least. True, it did rear up on its hind paws with a howl that started a landslide or two, and it did thrash the rocks with its tail until the sparks flew all over the canyon. But then it scratched its ear, cleared its throat and coolly continued on its way, though trotting at a slightly quicker pace. Unable to believe his eyes, Klapaucius ran along the ridge to head the creature off at the mouth of the dried-up stream—it was no longer an article, or even two articles in the *Dracological Journal* he could see his name on now, but a whole monograph elegantly bound, with a likeness of the dragon and the author on the cover!

At the first bend he crouched behind a boulder, pulled out his improbability automatic, took aim and actuated the possibilibalistic destabilizers. The gunstock trembled in his hands, the red-hot barrel steamed; the dragon was surrounded with a halo like a moon predicting bad weather—but didn't disappear! Once again Klapaucius unleashed the utmost improbability at the beast; the intensity of nonverisimilarity was so great, that a moth that happened to be flying by began to tap out the *Second Jungle Book* in Morse code with its little wings, and here and there among the crags and cliffs danced the shadows of witches, hags and harpies, while the sound of hoofbeats announced that somewhere in the vicinity there were centaurs gamboling, summoned into being by the awesome force of the improbability projector. But the dragon just sat there and yawned, leisurely scratching its shaggy neck with a hind paw, like a dog. Klapaucius clutched his sizzling weapon and desperately kept squeezing the trigger—he had never felt so helpless—and the nearest stones slowly lifted into the air, while the dust that the dragon had kicked up, instead of setting, hung in midair and assumed the shape of a sign that clearly read AT YOUR SERVICE GOV. It grew dim—day was night and night was day, it grew cold—hell was freezing over; a couple of stones went out for a stroll and softly chatted of this and that; in short, miracles were happening right and left, yet that horrid monster sitting not more than thirty paces from Klapaucius apparently had no intention of disappearing. Klapaucius threw down his gun, pulled an antidragon grenade from his vest pocket and, committing his soul to the Universal Matrix of Transfinite Transformations, hurled it with all his might. There was a loud ker-boom, and into the air with a spray of rock flew the dragon's tail, and the dragon shouted "Yipe!"—just like a person—and galloped straight for Klapaucius. Klapaucius, seeing the end was near, leaped out from behind his boulder, swinging his antimatter saber blindly, but then he heard another shout:

"Stop! Stop! Don't kill me!"

"What's that, the dragon talking?" thought Klapaucius. "I must be going mad . . ."

But he asked:

"Who said that? The dragon?"

"What dragon? It's me!!"

And as the cloud of dust blew away, Trurl stepped out of the beast, pushing a button that made it sink to its knees and go dead with a long, drawn-out wheeze.

"Trurl, what on earth is going on? Why this masquerade?

Where did you find such a costume? And what about the real dragon?" Klapaucius bombarded his friend with questions. Trurl finished brushing himself off and held up his hands.

"Just a minute, give me a chance! The dragon I destroyed, but the King wouldn't pay . . ."

"Why not?"

"Stingy, most likely. He blamed it on the bureaucracy, of course, said there had to be a notarized death certificate, an official autopsy, all sorts of forms in triplicate, the approval of the Royal Appropriations Commission, and so on. The Head Treasurer claimed he didn't know the procedure to hand over the money, for it wasn't wages, nor did it come under maintenance. I went from the King to the Cashier to the Commission, back and forth, and no one would do anything; finally, when they asked me to submit a vita sheet with photographs and references, I walked out—but by then the dragon was beyond recall. So I pulled the skin off it, cut up a few sticks and branches, found an old telephone pole, and that was really all I needed; a frame for the skin, some pulleys—you know—and I was ready . . ."

"You, Trurl? Resorting to such shameful tactics? Impossible! What could you hope to gain by it? I mean, if they didn't pay you in the first place . . ."

"Don't you understand?" said Trurl, shaking his head. "This way I get the tribute! Already there's more than I know what to do with."

"Ah! Of course!!" Klapaucius saw it all now. But he added, "Still, it wasn't right to force them . . ."

"Who was forcing them? I only walked around in the mountains, and in the evenings I howled a little. But really, I'm absolutely bushed." And he sat down next to Klapaucius.

"What, from howling?"

"Howling? What are you talking about? Every night I have to drag sacks of gold from the designated cave—all the way up there!" He pointed to a distant ridge. "I made myself a blast-off pad—it's right over there. Just carry several hundred pounds of bullion from sundown to sunup and you'll see what I mean! And that dragon was no ordinary dragon—the skin itself weighs a couple of tons, and I have to cart that around with me all day, roaring and stamping—and then it's all night hauling and heaving. I'm glad you showed up, I can't take much more of this. . . ."

"But . . . why didn't the dragon—the fake one, that is—why didn't it disappear when I lowered the probability to the point of miracles?" Klapaucius asked. Trurl smiled.

"I didn't want to take any chances," he explained. "Some fool of a hunter might've happened by, maybe even Basiliscus himself, so I put probability-proof shields under the dragon-skin. But come, I've got a few sacks of platinum left—saved them for last since they're the heaviest. Which is just perfect, now that you can give me a hand . . ."

Concerning Irregular Figures

Edwin Abbott Abbott

(from *Flatland*)

Flatland, first published in 1884, is the liberal theologian Edwin Abbott Abbott's satirical account of the manners and mores of Victorian England. Abbott draws on the non-Euclidean and higher-dimensional geometries developed by nineteenth-century mathematicians, which were part of the cutting-edge science of the day as well as subjects of great popular interest. In the decades following 1884 these ideas helped inspire H. G. Wells's *The Time Machine,* Einstein's relativity theory, and the cubist painting of Picasso and Braque.

Flatland's narrator, A Square, is an inhabitant of the two-dimensional country called Flatland, where people are geometrical figures and social standing is determined by the number of sides one possesses. Women are straight lines; soldiers and laborers are isosceles triangles (the narrower are the lower-class); tradesmen and shopkeepers are equilateral triangles; professionals and gentlemen are squares and pentagons; hexagons and higher are the nobility; and circles are priests. After spending some pages describing the interactions among these symmetrical figures, A Square turns to the asymmetrical. It is hard to read this excerpt without thinking of phrenology, the nineteenth-century "science" that tried to detect criminal and deviant tendencies through study of the shape of one's skull—or of eugenics, a term coined by Francis Galton just a year before *Flatland* appeared.

Throughout the previous pages I have been assuming—what perhaps should have been laid down at the beginning as a distinct and fundamental proposition—that every human being in Flatland is a Regular Figure, that is to say of regular construction. By this I mean that a Woman must not only be a line, but a straight line; that an Artisan or Soldier must have two of his sides equal; that Tradesmen must have three sides equal; Lawyers (of which class I am a humble member), four sides equal, and, generally, that in every Polygon, all the sides must be equal.

The size of the sides would of course depend upon the age of the individual. A Female at birth would be about an inch long, while a tall adult Woman might extend to a foot. As to the Males of every class, it may be roughly said that the length of an adult's sides, when added together, is two feet or a little more. But the size of our sides is not under consideration. I am speaking of the *equality* of sides, and it does not need much reflection to see that the whole of the social life in Flatland rests upon the fundamental fact that Nature wills all Figures to have their sides equal.

If our sides were unequal our angles might be unequal. Instead of its being sufficient to feel, or estimate by sight, a single angle in order to determine the form of an individual, it would be necessary to ascertain each angle by the experiment of Feeling. But life would be too short for such a tedious grouping. The whole science and art of Sight Recognition would at once perish; Feeling, so far as it is an art, would not long survive; intercourse would become perilous or impossible; there would be an end to all confidence, all forethought; no one would be safe in making the most simple social arrangements; in a word, civilization would relapse into barbarism.

Am I going too fast to carry my Readers with me to these obvious conclusions? Surely a moment's reflection, and a single instance from common life, must convince every one that our whole social system is based upon Regularity, or Equality of Angles. You meet, for example, two or three Tradesmen in the street, whom you recognize at once to be Tradesmen by a glance at their angles and rapidly bedimmed sides, and you ask them to step into your house to lunch. This you do at present with perfect confidence, because everyone knows to an inch or two the area occupied by an adult Triangle: but imagine that your Tradesman drags behind his regular and respectable vertex, a parallelogram of twelve or thirteen inches in diagonal:—what are you to do with such a monster sticking fast in your house door?

But I am insulting the intelligence of my Readers by accumulating details which must be patent to everyone who enjoys the advantages of a Residence in Spaceland. Obviously the measurements of a single angle would no longer be sufficient under such portentous circumstances; one's whole life would be taken up in feeling or surveying the perimeter of one's acquaintances. Already the difficulties of avoiding a collision in a crowd are enough to tax the sagacity of even a well-educated Square; but if no one could calculate the Regularity of a single figure in the company, all would be chaos and confusion, and the slightest panic would cause serious injuries, or—if there happened to be any Women or Soldiers present—perhaps considerable loss of life.

Expediency therefore concurs with Nature in stamping the seal of its approval upon Regularity of conformation: nor has the Law been backward in seconding their efforts. "Irregularity of Figure" means with us the same as, or more than, a combination of moral obliquity and criminality with you, and is treated accordingly. There are not wanting, it is true, some promulgators of paradoxes who maintain that there is no necessary connection between geometrical and moral Irregularity. "The Irregular," they say, "is from his birth scouted by his own parents, derided by his brothers and sisters, neglected by the domestics, scorned and suspected by society, and excluded from all posts of responsibility, trust, and useful activity. His every movement is jealously watched by the police till he comes of age and presents himself for inspection; then he is either destroyed, if he is found to exceed the fixed margin of deviation, or else immured in a Government Office as a clerk of the seventh class; prevented from marriage; forced to drudge at an uninteresting occupation for a miserable stipend; obliged to live and board at the office, and to take even his vacation under close supervision; what wonder that human nature, even in the best and purest, is embittered and perverted by such surroundings!"

All this very plausible reasoning does not convince me, as it has not convinced the wisest of our Statesmen, that our ancestors erred in laying it down as an axiom of policy that the toleration of Irregularity is incompatible with the safety of the State. Doubtless, the life of an Irregular is hard; but the interests of the Greater Number require that it shall be hard. If a man with a triangular front and a polygonal back were allowed to exist and to propagate a still more Irregular posterity, what would become of the arts of life? Are the houses and doors and churches in Flatland to be altered in order to accommodate such monsters? Are our ticket-collectors to be required to measure every man's perimeter before they allow him to enter a theatre, or to take his place in a lecture room? Is an Irregular to be exempted from the militia? And if not, how is he to be prevented from carrying desolation into the ranks of his comrades? Again, what irresistible temptations to fraudulent impostures must needs beset such a creature! How easy for him to enter a shop with his polygonal front foremost, and to order goods to any extent from a confiding tradesman! Let the advocates of a falsely called Philanthropy plead as they may for the abrogation of the Irregular Penal Laws, I for my part have never known an Irregular who was not also what Nature evidently intended him to be—a hypocrite, a misanthropist, and, up to the limits of his power, a perpetrator of all manner of mischief.

Not that I should be disposed to recommend (at present) the extreme measures adopted in some States, where an infant whose angle deviates by half a degree from the correct angularity is summarily destroyed at birth. Some of our highest and ablest men, men of real genius, have during their earliest days laboured under deviations as great as, or even greater than, forty-five minutes: and the loss of their precious lives would have been an irreparable injury to the State. The art of healing also has achieved some of its most glorious triumphs in the compressions, extensions, trepannings, colligations, and other surgical or diaetetic operations by which Irregularity has been partly or wholly cured. Advocating therefore a *Via Media,* I would lay down no fixed or absolute line of demarcation; but at the period when the frame is just beginning to set, and when the Medical Board has reported that recovery is improbable, I would suggest that the Irregular offspring be painlessly and mercifully consumed.

The Definition of Love

Andrew Marvell

My Love is of a birth as rare
As 'tis for object strange and high;
It was begotten by Despair
Upon Impossibility.

Magnanimous Despair alone
Could show me so divine a thing,
Where feeble Hope could ne'er have flown
But vainly flapped its tinsel wing.

And yet I quickly might arrive
Where my extended soul is fixt,
But Fate does iron wedges drive,
And always crowds itself betwixt.

For Fate with jealous eye does see
Two perfect Loves; nor lets them close:
Their union would her ruin be,
And her tyrannic power depose.

And therefore her decrees of steel
Us as the distant Poles have placed,
(Though Love's whole World on us doth wheel)
Not by themselves to be embraced.

Unless the giddy Heaven fall,
And Earth some new convulsion tear;
And, us to join, the World should all
Be cramped into a planisphere.

As lines so Loves oblique may well
Themselves in every angle greet:
But ours so truly parallel,
Though infinite, can never meet.

Therefore the Love which us doth bind
But Fate so enviously debars,
Is the conjunction of the Mind,
And opposition of the Stars.

On Fiddib Har

A. K. Dewdney

(From *The Planiverse: Computer Contact with a Two-Dimensional World*)

The Planiverse originates in the late 1970s as a class project among the narrator's computer science students at the University of Western Ontario (where A. K. Dewdney taught computer science until his recent retirement). Originally designed only to model simple collisions between particles in a two-dimensional universe, the project gradually becomes more and more complex, acquiring physical laws, stars, planets, and finally life. One year several undergraduates decide to stay through the summer to work on the model: a biology major wishes to design more complex life-forms for it; a computer science major creates a "flexible, multipurpose query system" enabling one to communicate with the program while it is running; another computer science student devises a way to make the program much more elaborate while permitting it to run even faster than before. The result is that, like a Julia set in one of Robert Devaney's fractal movies, the Planiverse "explodes" into a new level of complexity.

One day the program's responses change. It uses words that have not been programmed into it, and begins to ask questions. Dewdney and his students realize that their computer universe has spawned a civilization of sentient creatures.

The group discovers that their primary contact is a young person living on a planet called Arde, on the continent of Ajem Kollosh. Most of *The Planiverse* is concerned with describing this two-dimensional planet and the civilization that has arisen on it, as seen through the eyes and philosophy of the young wanderer, Yendred. The planet's one ocean is called Fiddib Har, and it is there that Yendred and his father, as the story opens, make their living as fishermen.

Friday, May 30, 2:00 P.M.

It was nearly two Arde-days before Yendred was to leave on a journey that would span half the continent of Ajem Kollosh. When he came into focus on our screen, Yendred was walking in front of his father toward the shore of Fiddib Har. According to our calculations it was nearly midnight. The two walked without speaking because Yendred was communicating with us much of this time. He had agreed to help his father for the next day. By midnight a strong east wind would be blowing them far out on Fiddib Har, where they would spend the morning fishing. They would then sail back with the afternoon westerly.

As they walked, they passed over the five houses between Yendred's house and the sea. All of these, as well as most of the houses on the other side of Yendred's, were inhabited by fishing families. On Arde, communities tend to form around occupations not only because so much short-distance travel is on foot but because the simple linear geography makes the advantages of living close to one's work glaringly obvious.

Following the progress of Yendred and his father, we became aware that they were part of a procession moving toward the shore. These were fishermen who had all left their homes at the same time. Their boats were stored ahead of them in the same order as their houses, and all would arrive, more or less simultaneously, at their boathouses, no one having to climb over anyone else to get there. Not surprisingly, the boathouses were underground sheds. These lay beneath the sandy kilometer of gently sloping beach between the seashore and the nearest house.

There seemed to be an air of eagerness about the fishermen.

I could imagine the brilliant, diamond-studded night sky, its bright glow reflected on the sand, a slowly freshening offshore breeze, and the cheerful banter of beings simplified by the sea. So it was that Yendred and his father arrived at the sixth shed from the water and proceeded to exhume its roof, sliding it toward the shore so that it became continuous with the sloping shed floor. Yendred's father crossed the floor, climbed over the boat, and planted himself behind its stern. Yendred reached into the boat and withdrew a roller disk, which he wedged under the bow. Pushing and pulling, the two rolled the boat up out of the shed and over the roof. His father retrieved the rollers as they appeared from below the stern, tossing them forward to bounce along the horizontal mast to Yendred, who would catch them and plant them once more under the bow. They halted this process only once—to replace the shed roof—and then continued to the shoreline, where the boat slid easily into the calm water.

When the mast had been put up and all the gear and tackle made ready, we realized that the boat had no bow or stern as such. Being completely symmetrical about its mast, bow and stern could be determined only by which direction the boat was moving. This ceased to puzzle us when we realized that a two-dimensional boat could not "turn around."

The mast of the boat has two sections, a stout lower section inserted in a well-braced framework, and a graceful upper section resting on this and held in place by two wedges or pegs. Two long ropes dangled from the top of the mast. There were lockers for food, tackle, and line, as well as two holds for the catch. At either end of the boat were oars. One of these was soon being pumped by Yendred's father; the other was folded up to be used by Yendred on their return. Ingeniously designed, the blades of the oars were hinged so that when the rower pulled down on the handle and the blades stroked toward the rear, they formed a rigid paddle. When the rower pushed up on the handle, the blades moved forward, folding out of the way so as not to counteract the boat's forward motion. There were no visible fasteners used in the boat's construction; it was all held together by glue.

Although it was night, things appeared on our screen just

as they would in the daytime. The sea bottom was barely two meters below the boat and, judging from the boat's progress over colonies of strange shellfish dwelling there, it was now moving forward swiftly. Yendred's father ceased rowing and folded up the oar, securing it with a rope that ran from the tip of the bottom blade up to a peg which he inserted in a slot near his feet. The easterly was clearly strengthening; it blew directly upon the mast, this being all a two-dimensional boat requires by way of a sail.

We asked Yendred about the shellfish which his boat had passed over.

- YOU PASSED OVER SOME ANIMALS OR PLANTS LYING ON THE SEA BOTTOM. WHAT ARE THEY?

  BALAT SRAR. ANIMALS IN SHELLS WHICH CAN OPEN AND CLOSE. WE EAT THEM.

- WE ARE GUESSING THAT THEY EAT TINY FOOD PARTICLES. IS THIS TRUE?

  YES. DO YOU ALSO HAVE BALAT SRAR AT EARTH?

- SOMETHING SIMILAR. HOW DO THEY OPEN AND CLOSE?

THEY PUMP WATER INTO THE HINGE OF THEIR SHELL AND FORCE IT OPEN. TO CLOSE THEY WATER PUMP OUT AGAIN.

Since Yendred and his father were on opposite sides of the mast, he could continue this conversation undistracted.

Although the Balat Srar have similarities to our clams, it would be more accurate to compare them with lamp shells or brachiopods, once very common on Earth.

The Balat Srar lives between two curved valves which are hinged at one end. Attached to these valves by a mantle, it consists of fleshy feeding lobes, a "pump muscle," and a pump chamber next to the hinge. A primitive nervous system directs the activities of the Balat Srar. Although clams and lamp shells here on Earth open and close by muscles near their hinge, the Balat Srar does not do so directly. For example, merely contracting its pump muscle has no result since the water in its chamber cannot be compressed. Nevertheless, by employing its zipper organ, it can easily pump water into or out of the chamber, opening and closing its shell.

Watching the Balat Srar opening and closing earlier, and now hearing Yendred's description of their muscular pump, a puzzling

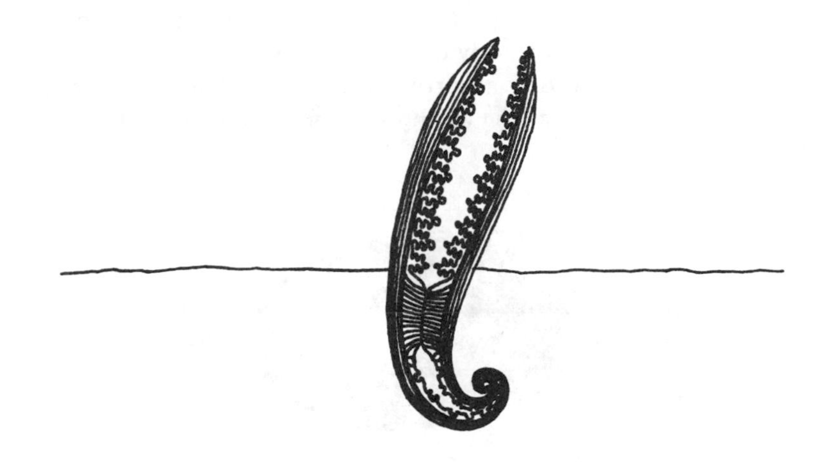

feature of Ardean anatomy suddenly became clear: Ardeans move their arms and legs in precisely the same way that this humble shellfish opens and closes its valves. Ardean muscles do not so much contract and expand as they pump fluids into and out of the chamber at each joint. In fact some joints in the arms have two chambers separated by an articulated bone. This gives Ardean arms great flexibility of movement.

Yendred was very enthusiastic on the subject of biology and seemed to know a great deal about the Balat Srar. He explained how the Balat Srar grows by secreting a new layer of shell each year. When fully mature, the male Balat Srar releases sperm into the water at a certain time of the year. Nearby females, which begin rhythmically to open and close their shells, eventually take the sperm into their chambers, lined with unfertilized eggs. When fertilized, the eggs develop into larvae, become detached from the chamber wall, and are expelled into the water of Fiddib Har. At this point, most of the larvae are eaten, but a few survive.

We learned about these larvae much later in Yendred's journey, when we came across the Ardean equivalent of a biological laboratory and had the opportunity to examine the larvae closely and have them explained (through Yendred) by a Punizlan scientist. Apparently, the pump muscle is active from a Balat Srar's earliest moments. It comprises 80 percent of the larva's body mass and, in concert with a smaller, temporary muscle at its head, the larva swims by a kind of jet propulsion, feeding on the organic particles which enter its body cavity in the process.

The young Balat Srar absorbs food into its digestive lobes in exactly the same way as the adult, and the food diffuses through

the lobes by a rather interesting mechanism. When the larva reaches a certain point in its development, it ceases to swim, sinks to the sea bottom, and grows two hard shells which spiral as they grow, at first sharply and then more gradually. The spirals meet at one end and form a hinge, which always wears away enough to prevent locking of the valves.

During our discussion of the Balat Srar, the wind was growing steadily stronger and, all too soon, Yendred had to break off in order to tell his father (and us) that they were approaching the boat ahead of them in the fishing fleet. Soon their mast would rob all the wind from the mast of the boat ahead and they would ram it. Should Yendred take out the mast peg or deploy the forward oar in order to slow their craft?

> NOW HE SAYS TO TAKE OUT THE PEG. HE HOLDS THE MAST ROPE ON THE OTHER SIDE. IT IS HARD TO HEAR HIM IN THIS WIND. I HADD USE TO HIT THE PEG. IT IS COMING LOOSE. I PULL IT FROM THE SLOT AND HOLD IT AGAINST THE MAST. MY FATHER LETS THE MAST SWING DOWN OVER ME WITH THE ROPE. HE ASKS IF WE ARE SLOWING. YES BUT WAIT. WAIT. NOW PULL UP THE MAST. HE IS VERY STRONG TO DO THAT IN A WIND. NOW THE MAST UP AGAIN IS AND I HIT THE PEG INTO THE SLOT AGAIN. THE OTHER BOAT IS VERY AHEAD AGAIN.

[Hadd is a kind of Ardean metal.]

In the course of their voyage they had to repeat this maneuver two more times. The wind grew steadily stronger and Yendred's father moved toward the (temporary) stern, occasionally to pull on the rope in order to decrease wind resistance and their resulting speed. The danger was that a sudden gust would cause the mast to break or the boat to pitchpole. As it was, a great mound of water had piled up ahead of the boat as it plowed through the sea. Occasionally, they came upon floating water plants. These would circulate for a while within the bow wave and then slip under the boat. We let Yendred's boat pass off the screen in order to focus upon one of these. It was not difficult to see how it would look when the sea was calm.

The "Ilma Kabosh," as Yendred called it, has no stalk, merely a

sort of root organ joined directly to a symmetrical arrangement of leaves, four on each side. The root organ absorbs minerals and other nutrients from the water through thousands of tiny hairs covering its branched lobes. The bottom pair of leaves produce a gas which fills the buoyancy chambers formed by the second pair of leaves. These curl around and stick to the undersurface of the bottom pair. The third pair of leaves rest upon the buoyancy chambers. These are filled with cells containing "hadrashar," which, according to Yendred, enables the plant to manufacture complex foods from shemlight and the rather simple nutrients it absorbs through its roots. The fourth set of leaves do the same thing except that they can also be moved into an upright position to act as sails. At the very top of the plant is a central arrangement of reproductive organs, a single egg stalk surrounded by two pollinators.

Eventually the wind began to die down and, by early morning, it had become calm enough to fish. Yendred took down the mast, resting it upon the water beside him, and fed line to his father from a locker at the foot of the mast. His father attached some rather vicious-looking traps to the line at intervals and fed these out, over his head, into the water below.

The fish trap consisted of two jaws, set with teeth and held open by a pin placed inside the trap. A morsel of bait stuck to the pin was all that was needed to lure their catch into the trap. As they set the traps and put them overboard, one by one, we tried to imagine what sort of creature they hoped to catch.

The clock on the laboratory wall was nearing midnight our time, even as Yendred and his father enjoyed their midmorning calm. Alice sat at the terminal staring into the screen. Lambert suggested that she follow the line down to see what Ardean swimmer took the bait.

With a sigh, she punched some tracking data into the key-

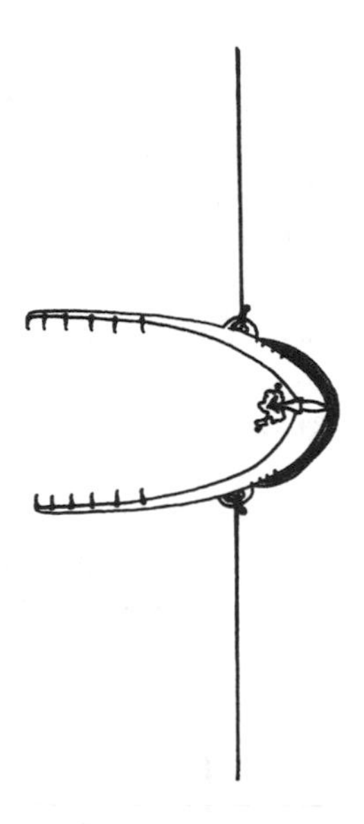

board and we began to "descend" along the line. Every meter or so there was a tassel attached to the line. Yendred's father had used these to grip the line, as Ardean hands (or *any* two-dimensional hands) cannot grasp a line as ours can.

Alice stopped the scan when she arrived at the first trap: we could not remove ourselves very far from Yendred's current location without threatening the Earth/Arde link. Even as we watched this trap, it began to jerk slightly as the trap below it appeared to have caught something.

"Scan down. Scan down!" Lambert was excited.

"Be patient. I'm sure . . ."

Before she had even finished her sentence, a rather horrible-looking thing swam into view, investigating the trap. Luckily (not for it), the creature was on the proper side of the line to be caught. It looked like a cross between a centipede and a fish. Nosing into the trap, it suddenly seized the bait and began to worry it, gradually loosening the pin. When the pin suddenly popped free, the jaws of the trap closed with frightening swiftness, expelling the water and almost expelling the centipede-fish along with it. But the teeth of the trap caught the creature's fins and held it fast as it struggled up and down. What a curious contrast there was between the beautifully coordinated motions of this organism and the ugly reality which it now faced. It is like the way we feel as children when we see a helpless animal dying. Here was a completely new animal, one whose death I was not inured to.

Abruptly, the trap was pulled upward and we watched as the lower traps glided, one by one, into view. Most of them held fish, which were chiefly of two kinds. Yendred called them "Ara Hoot" and "Kobor Hoot."

Generally speaking, the Ara Hoot is the smaller of the two fish. It has from four to eight segments, each consisting of a pair of bony outer plates, transverse muscle with a central zipper, and

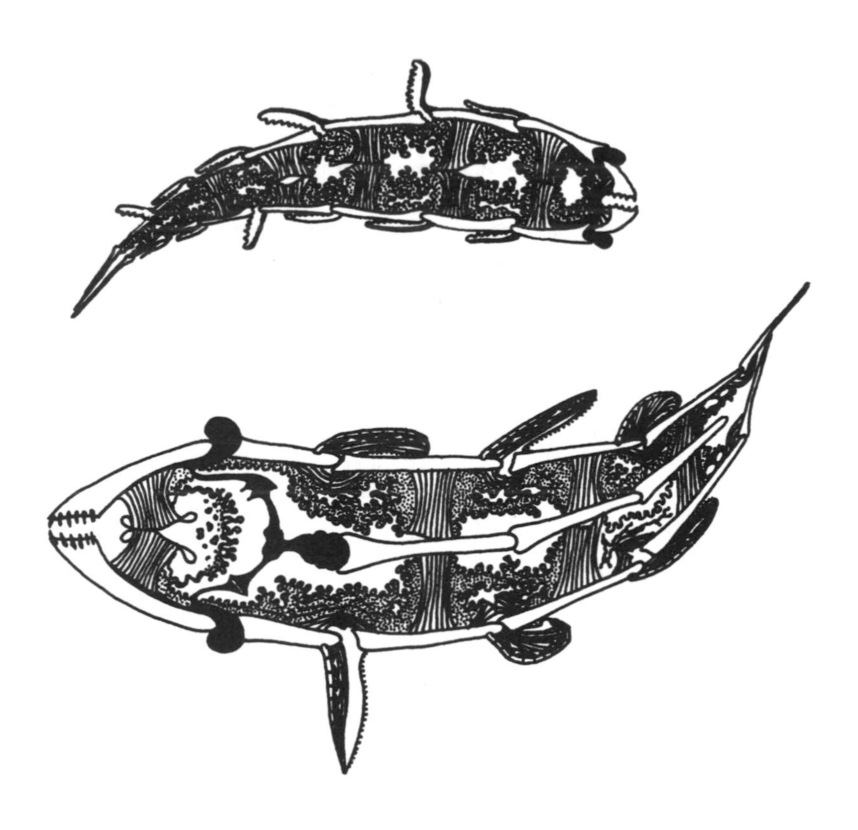

digestive/respiratory tissue. At the head is a pair of hard, transparent eyes upon which the jaws hinge, touch and taste organs and a very simple brain in two halves. The two halves communicate by a nerve trunk across the trailing edge of the jaw muscle, with the result that every time the creature swallows something this communication is temporarily interrupted. Nerve trunks also run from the brain halves toward the rear of the fish, following the inner surface of the bones as well as penetrating the muscles. Evidently, nerve tissue is not only able to transmit impulses but also has structural strength.

Food captured by the Ara Hoot is crushed into tiny pieces between the jaws and passed into the body cavity of the first segment. Here, and in subsequent segments, digestive tissue actively probes the body cavity for these particles, much like the Balat Srar digestive lobes. Ultimately, whatever remains, along with digestive wastes, is passed out of the final segment between two tailbones which are used for steering the creature.

The Ara Hoot swims by sweeping backward with extended fins. At the end of each stroke the fins retract, the fluid being sucked out of them by the zippered muscles which connect the nearby external bones. In this way, they offer little resistance to the water and sudden reinflation by the same zippers causes them to stand out once more from the body, ready for another rearward stroke.

The Kobor Hoot is much more advanced, anatomically speaking, than its smaller colleague. The chief difference lies in the Kobor Hoot's internal circulation of body fluid made possible by a series of jointed, semi-rigid "bones" running down the middle of its three body segments. The Kobor Hoot also has a larger, integrated brain attached to the foremost rod and communicating with the rest of the body by nerve trunks. One of the chief coordinating tasks of the brain is to adjust the pressure in each cavity so that exterior bones are forced open at the appropriate times to allow passage of fluids into or out of the "fins."

At the tail of the Kobor Hoot, within the last segment, are the waste-filtering and reproductive organs. Although the Ara Hoot expels solid wastes (in particle form), the Kobor Hoot has only liquid waste to get rid of. This is filtered out of the body fluid and periodically released between the tailbones. It is in the last cavity that eggs or sperm are created, and the Kobor Hoot is the only Ardean animal, to our knowledge, capable of direct insemination; the male simply places its two steering bones between the female's and releases sperm. The young Kobor Hoot are nurtured in the tail cavity until developed enough to swim alongside the mother.

As his father pulled in the line and emptied the traps, he pitched the fish across the deck to Yendred, who lifted the cover of the hold and threw them in, one by one. None of them appeared very damaged by the traps and many continued to work their fins, eeling their way forward and back among their fellow creatures within the hold.

The traps were reset and rebaited, and the line was cast back into the sea. The surface of Fiddib Har was absolutely level and calm. Yendred was discussing with his father the long journey he would begin on the morrow. After a time his father fell silent and cocked his head to the east, apparently watching the fishermen in the next boat. Yendred used this opportunity to take up the matter of dimensions with us once again. It was Edwards who spoke with him, dictating to Alice at the terminal.

IN WHICH DIRECTION ARE YOU?

- YOU CANNOT POINT IN OUR DIRECTION. IF YOU COULD. YOUR ARM WOULD DISAPPEAR.

WHAT IS DIMENSION?

- HERE IS AN EXAMPLE. THE OCEAN BELOW YOU HAS TWO DIMENSIONS. THE SURFACE OF FIDDIB HAR HAS ONE DIMENSION. A POINT ON THAT SURFACE HAS NO DIMENSIONS. YOUR WHOLE WORLD, YOUR UNIVERSE, IS TWO-DIMENSIONAL, LIKE FIDDIB HAR.

DOES OUR UNIVERSE HAVE A SURFACE?

- NO. YOUR UNIVERSE HAS NO SURFACE.

THEN HOW IT LIKE FIDDIB HAR IS?

Swimming and Breathing

The swimming fin or "oargan" of the Kobor Hoot consists of a series of jointed bones on its leading edge, a complicated series of muscles next to this, a tissue layer of unknown purpose, a long, narrow cavity, and a gill on the trailing edge.

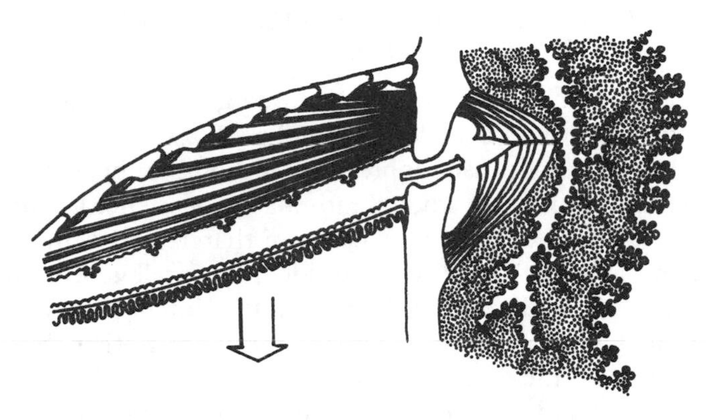

For the Kobor Hoot (and, for the matter, the Ara Hoot), to swim is to breathe. When the fin is inflated, it stands out from the body and the long muscles immediately contract, forcing the fin rapidly toward the rear. At this point in the swimming cycle, the gill at the rear of the fin is most actively collecting hrabx from the surrounding water. Simultaneously, the two body bones are disarticulated and the contracting space inside the fin pushes its fluid through the opening and against the portal muscle. The zipper of this muscle pumps all the fluid into the "pulmonary" cavity even as the bones come once again together and the short muscles of the fin begin their contraction phase. The fin begins to curl as it is drawn forward alongside the body. Meanwhile, having just inflated the pulmonary chamber, the portal muscle now deflates it, creating a bubble against the still articulated bones. When these are forced apart once more, the portal muscle contracts violently, inflating the fin again.

[We paused here, arguing among ourselves how to present this idea.]

- WHEN YOU LOOK DOWN AT THE SURFACE OF THE WATER, IMAGINE THAT CREATURES LIKE THREADS LIVED THERE. THEY WOULD NOT GO DOWN, BUT STAY ALWAYS AT THE SURFACE.

  IS THAT A WORLD OF ONE DIMENSION?

- YES. YOU LOOK AT THAT SPACE FROM OUTSIDE THAT SPACE. THIS IS HOW WE LOOK AT YOU.

Through such conversations, Yendred gained considerable insight into the third dimension, insight which we three-dimensional beings cannot easily appreciate. Although it helped considerably that Yendred had studied rather a lot of science in Punizlan schools, it was clear to us very early on that Yendred, by human standards at least, possessed considerable intelligence. At the same time, we all developed the feeling that intelligence is itself a universal thing, taking one material form or another, but ever striving toward the same goal.

Yendred's father hauled the line in, removed the fish, rebaited the traps, and let them down a third time. The eastern hold was filling with fish and the noon breeze was beginning from the west. We scanned eastward along the fishing fleet and found some of the crews beginning to put up their masts and stow the last catch. When we returned to Yendred's boat, we found Yendred installing the lower mast and his father slowly, almost with reluctance, hauling up the line for the last time.

Now Yendred put their fish, mostly Ara Hoots, in the western hold on his side of the boat. Finally, he took the line from his father and stuffed it all, nearly two hundred meters, into the line box. In doing this there was, of course, no danger that it would tangle. Before long the mast was up and the boat was moving slowly east. Every now and then the boat shuddered as a gust of wind caught its mast. We had noticed this phenomenon frequently during the night voyage as well. It was not until much later, when we could no longer contact Yendred, that we had the leisure to speculate about turbulence in a two-dimensional atmosphere.

Presently, the boat was ploughing eastward through the water, Yendred and his father having exchanged roles. Now it was Yendred's turn to hold the mast rope and spill air whenever it became necessary. Although the winds of Arde always come at the same time and from the same direction with almost perfect regularity, their force can vary considerably. On this occasion, the wind was almost visible as the mast quivered and the boat

TURBULENCE

Turbulence in air is a rather complicated flow in many different directions. Often, turbulence is composed of vortices or currents of air following a circular pattern, usually in the wake of some obstacle past which the air is moving. In three dimensions, the turbulence dissolves, so to speak, when its large vortices break up into smaller ones.

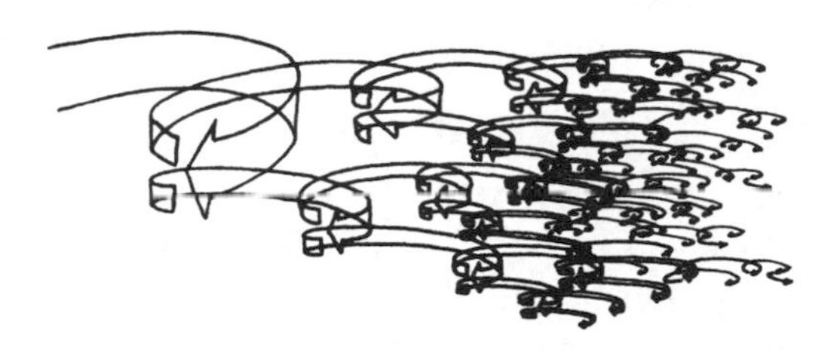

In two dimensions, turbulence does not dissipate nearly so readily. The vortices continue to propagate themselves for considerable distances along the wake of an obstacle.

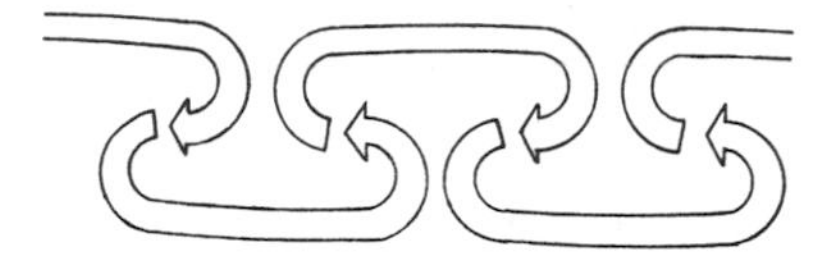

bounded over the foam. Water was collecting in the forward end of the boat and we feared that it might be swamped until we saw Yendred's father drag his lower, eastern arm along the floor of the boat, trapping all the water ahead of it. Following the inner contour of the boat, he literally swept all the water up the stern, over the oar, and out into the churning sea.

Suddenly, to our horror, the mast snapped just above the two wedges. Yendred had given the rope too much play and now it zipped from between his fingers and the mast struck his father's head, falling to the deck and leaning against his still upright body at a crazy angle.

Had Yendred's father been human, the force of the blow would undoubtedly have knocked him unconscious. Luckily, Ardean brains are not in Ardean heads and his father now stood his ground, waiting for Yendred to clamber over the stump of the mast and, balancing in the gale, to jump into his father's side of the boat.

Together they removed the stub of the old mast, pulled out the wedges, and replaced the now shortened mast in its hole. Behind them, to the west, five boats were spilling wind and plying their eastern oars, trying not to ram Yendred's boat or one another.

The six boats, with Yendred's in the lead, now made their way east in the gale at somewhat reduced speed. Yendred held the mast rope as firmly as he could, his two fingers surely aching.

As the sturdy craft bounded over the water of Fiddib Har, we zoomed in on the father's head to examine it closely. There appeared to be no damage either to his eye or jaw. Both were constructed of hard material. But the tissue below the eye on his father's western (left) side seemed to be displaced slightly. Would this not cause him pain? We wondered how Punizlan doctors (if there were any) might repair such damage. Surgery would certainly be a very different affair on Arde than it was on Earth. For one thing, cutting would not be done by a knife but by a needle. For another, suturing was impossible. We were to discover much later on that very little surgery is done on Arde, that most external wounds are treated with clamps and that the greater part of Punizlan medical practice was based on orally administered drugs.

By late afternoon the wind was dying and the last of the fishing fleet arrived at the beach. One by one the boats were rolled up on the firm sand, but they could not be as easily moved as on the previous night. The wind was now onshore and great breakers, the remnant of the afternoon westerly, broke against stumbling Ardean forms. Even so, the fishermen appeared to be disembarking much farther up the beach than where they had earlier set out. Without a moon, Arde could have no lunar tides. Yendred explained that this high water level was due entirely to the pressure of the west wind acting upon Fiddib Har. Even as the two pulled their boat up on the beach, one could see the water slowly recede.

Before they had gone a few meters Yendred and his father encountered an obstacle. A giant carnivorous plant called a Mil Dwili had washed up on the beach.

It is not hard to imagine how this plant must look when floating on the surface of Fiddib Har (see next page). It stays afloat by means of several buoyancy chambers and traps large fish between its two dangling arms. A system of roots then grows directly into the prey, absorbing its organic material. Was this an animal or a plant?

Yendred did not give us a clear answer. Since the boat could not be made to roll over the plant, his father picked it up and placed it under the mast. They rolled the boat up the gentle beach towards the shed, Yendred throwing the rollers to his father as they emerged from beneath the boat. Stowing the boat in the shed, they removed the day's catch and divided it up into two string bags. Then they replaced the shed roof, leaving the Mil Dwili to rot in the sand.

This was to be the longest period of continuous contact with Yendred we would ever experience. For almost 24 hours we had followed the fishing expedition. Outside the laboratory window burned the early afternoon sun. We were exhausted.

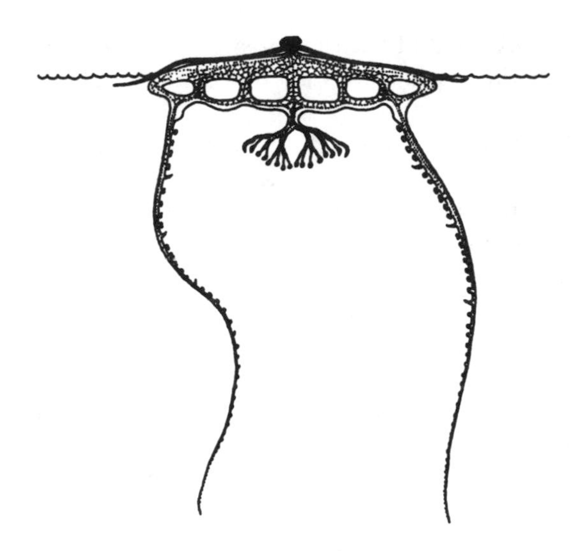

And so were Yendred and his father. They walked wearily home, each carrying a bag of fish. His father complained once about the soreness in his eye but said nothing else to Yendred. Yendred said nothing to us. Soon the sun would set in the west.

The Church of the Fourth Dimension

Martin Gardner

"Could I but rotate my arm out of the limits set to it," one of the Utopians had said to him, "I could thrust it into a thousand dimensions."

—H. G. Wells, *Men Like Gods*

Alexander Pope once described London as a "dear, droll, distracting town." Who would disagree? Even with respect to recreational mathematics, I have yet to make an imaginary visit to London without coming on something quite extraordinary. Last fall, for instance, I was reading the London *Times* in my hotel room a few blocks from Piccadilly Circus when a small advertisement caught my eye:

> WEARY OF THE WORLD OF THREE DIMENSIONS? COME WORSHIP WITH US SUNDAY AT THE CHURCH OF THE FOURTH DIMENSION. SERVICES PROMPTLY AT 11 A.M., IN PLATO'S GROTTO. REVEREND ARTHUR SLADE, MINISTER.

An address was given. I tore out the advertisement, and on the following Sunday morning rode the Underground to a station within walking distance of the church. There was a damp chill in the air and a light mist was drifting in from the sea. I turned the last corner, completely unprepared for the strange edifice that loomed ahead of me. Four enormous cubes were stacked in one column, with four cantilevered cubes jutting in four directions from the exposed faces of the third cube from the ground. I recognized the structure at once as an unfolded hypercube. Just as the six square faces of a cube can be cut along seven lines and unfolded to make a two-dimensional Latin cross (a popular floor plan for medieval churches), so the eight cubical hyperfaces of a four-dimensional cube can be cut along seventeen squares and "unfolded" to form a three-dimensional Latin cross.

A smiling young woman standing inside the portal directed me to a stairway. It spiraled down into a basement auditorium that I can only describe as a motion-picture theater combined with a limestone cavern. The front wall was a solid expanse of white. Formations of translucent pink stalactites glowed brightly on the ceiling, flooding the grotto with a rosy light. Huge stalagmites surrounded the room at the sides and back. Electronic organ music, like the score of a science-fiction film, surged into the room from all directions. I touched one of the stalagmites. It vibrated beneath my fingers like the cold key of a stone xylophone.

The strange music continued for ten minutes or more after I had taken a seat, then slowly softened as the overhead light began to dim. At the same time I became aware of a source of bluish light at the rear of the grotto. It grew more intense, casting sharp shadows of the heads of the congregation on the lower part of the white wall ahead. I turned around and saw an almost blinding point of light that appeared to come from an enormous distance.

The music faded into silence as the grotto became completely dark except for the brilliantly illuminated front wall. The shadow of the minister rose before us. After announcing the text as Ephesians, Chapter 3, verses 17 and 18, he began to read in low, resonant tones that seemed to come directly from the shadow's head: ". . . that ye, being rooted and grounded in love, may be able to comprehend with all saints what is the breadth, and length, and depth, and height. . . ."

It was too dark for note-taking, but the following paragraphs summarize accurately, I think, the burden of Slade's remarkable sermon.

Our cosmos—the world we see, hear, feel—is the three-dimensional "surface" of a vast, four-dimensional sea. The ability to visualize, to comprehend intuitively, this "wholly other" world of higher space is given in each century only to a few chosen seers. For the rest of us, we must approach hyperspace indirectly, by way of analogy. Imagine a Flatland, a shadow world of two dimensions like the shadows on the wall of Plato's famous cave (*Republic,* Chapter 7). But shadows do not have material substance, so it is best to think of Flatland as possessing an infinitesimal thickness equal to the diameter of one of its fundamental particles. Imagine these particles floating on the smooth surface of a liquid. They dance in obedience to two-dimensional laws. The inhabitants of Flatland, who are made up of these particles, cannot conceive of a third direction perpendicular to the two they know.

We, however, who live in three-space can see every particle of Flatland. We see inside its houses, inside the bodies of every Flatlander. We can touch every particle of their world without passing our finger through their space. If we lift a Flatlander out of a locked room, it seems to him a miracle.

In an analogous way, Slade continued, our world of three-space floats on the quiet surface of a gigantic hyperocean; perhaps, as Einstein once suggested, on an immense hypersphere. The four-dimensional thickness of our world is approximately the diameter of a fundamental particle. The laws of our world are the "surface tensions" of the hypersea. The surface of this sea is uniform, otherwise our laws would not be uniform. A slight curvature of the sea's surface accounts for the slight, constant curvature of our space-time. Time exists also in hyperspace. If time is regarded as our fourth coordinate, then the hyperworld is a world of five dimensions. Electromagnetic waves are vibrations on the surface of the hypersea. Only in this way, Slade emphasized, can science escape the paradox of an empty space capable of transmitting energy.

What lies outside the sea's surface? The wholly other world of God! No longer is theology embarrassed by the contradiction between God's immanence and transcendence. Hyperspace touches every point of three-space. God is closer to us than our breathing. He can see every portion of our world, touch every particle without moving a finger through our space. Yet the Kingdom of God is completely "outside" of three-space, in a direction in which we cannot even point.

The cosmos was created billions of years ago when God poured (Slade paused to say that he spoke metaphorically) on the surface of the hypersea an enormous quantity of hyperparticles with asymmetric three-dimensional cross sections. Some of these particles fell into three-space in right-handed form to become neutrons, the others in left-handed form to become antineutrons. Pairs of opposite parity annihilated each other in a great primeval explosion, but a slight preponderance of hyperparticles happened to fall as neutrons and this excess remained. Most of these neutrons split into protons and electrons to form hydrogen. So began the evolution of our "one-sided" material world. The explosion caused a spreading of particles. To maintain this expanding universe in a reasonably steady state, God renews its matter at intervals by dipping his fingers into his supply of hyperparticles and flicking them toward the sea. Those which fall as antineutrons are annihilated, those which fall as neutrons remain. Whenever an antiparticle is created in the laboratory, we witness an actual "turning over" of an asymmetric particle in the same way that one can reverse in three-space an asymmetric two-dimensional pattern of cardboard. Thus the production of antiparticles provides an empirical proof of the reality of four-space.

Slade brought his sermon to a close by reading from the recently discovered Gnostic Gospel of Thomas: "If those who lead you say to you: Behold the kingdom is in heaven, then the birds will precede you. If they say to you that it is in the sea, then the fish will precede you. But the kingdom is within you and it is outside of you."

Again the unearthly organ music. The blue light vanished, plunging the cavern into total blackness. Slowly the pink stalactites overhead began to glow, and I blinked my eyes, dazzled to find myself back in three-space.

Slade, a tall man with iron-gray hair and a small dark mustache, was standing at the grotto's entrance to greet the members of his congregation. As we shook hands I introduced myself and mentioned this department. "Of course!" he exclaimed. "I have some of your books. Are you in a hurry? If you wait a bit, we'll have a chance to chat."

After the last handshake Slade led me to a second spiral stairway of opposite handedness from the one on which I had descended earlier. It carried us to the pastor's study in the top cube of the church. Elaborate models, three-space projections of various types of hyperstructures, were on display around the room. On one wall hung a large reproduction of Salvador Dali's painting "Corpus Hypercubus." In the picture, above a flat surface of checkered squares, floats a three-dimensional cross of eight cubes; an unfolded hypercube identical in structure with the church in which I was standing.

"Tell me, Slade," I said, after we were seated, "is this doctrine of yours new or are you continuing a long tradition?"

"It's by no means new," he replied, "though I can claim to have established the first church in which hyperfaith serves as the cornerstone. Plato, of course, had no conception of a geometrical fourth dimension, though his cave analogy clearly implies it. In fact, every form of Platonic dualism that divides existence into the natural and supernatural is clearly a nonmathematical way of speaking about higher space. Henry More, the seventeenth-century Cambridge Platonist, was the first to regard the spiritual world as having four spatial dimensions. Then along came Immanuel Kant, with his recognition of our space and time as subjective lenses, so to speak, through which we view only a thin slice of transcendent reality. After that it is easy to see how the concept of higher space provided a much needed link between modern science and traditional religions."

"You say 'religions,' " I put in. "Does that mean your church is not Christian?"

"Only in the sense that we find essential truth in all the great world faiths. I should add that in recent decades the Continental Protestant theologians have finally discovered four-space. When

Karl Barth talks about the 'vertical' or 'perpendicular' dimension, he clearly means it in a four-dimensional sense. And of course in the theology of Karl Heim there is a full, explicit recognition of the role of higher space."

"Yes," I said. "I recently read an interesting book called *Physicist and Christian,* by William G. Pollard [executive director of the Oak Ridge Institute of Nuclear Studies and an Episcopal clergyman]. He draws heavily on Heim's concept of hyperspace."

Slade scribbled the book's title on a note pad. "I must look it up. I wonder if Pollard realizes that a number of late-nineteenth-century Protestants wrote books about the fourth dimension. A. T. Schofield's *Another World,* for example [it appeared in 1888], and Arthur Willink's *The World of the Unseen* [subtitled "An Essay on the Relation of Higher Space to Things Eternal"; published in 1893]. Of course modern occultists and spiritualists have had a field day with the notion. Peter D. Ouspensky, for instance, has a lot to say about it in his books, although most of his opinions derive from the speculations of Charles Howard Hinton, an American mathematician. Whately Carington, the English parapsychologist, wrote an unusual book in 1920—he published it under the by-line of W. Whately Smith—on *A Theory of the Mechanism of Survival.*"

"Survival after death?"

Slade nodded. "I can't go along with Carington's belief in such things as table tipping being accomplished by an invisible four-dimensional lever, or clairvoyance as perception from a point in higher space, but I regard his basic hypothesis as sound. Our bodies are simply three-dimensional cross sections of our higher four-dimensional selves. Obviously a man is subject to all the laws of this world, but at the same time his experiences are permanently recorded—stored as information, so to speak—in the four-space portion of his higher self. When his three-space body ceases to function, the permanent record remains until it can be attached to a new body for a new cycle of life in some other three-space continuum."

"I like that," I said. "It explains the complete dependence of mind on body in this world, at the same time permitting an unbroken continuity between this life and the next. Isn't this close to what William James struggled to say in his little book on immortality?"

"Precisely. James, unfortunately, was no mathematician, so he had to express his meaning in nongeometrical metaphors."

"What about the so-called demonstrations of the fourth dimension by certain mediums," I asked. "Wasn't there a professor of astrophysics in Leipzig who wrote a book about them?"

I thought I detected an embarrassed note in Slade's laugh. "Yes, that was poor Johann Karl Friedrich Zöllner. His book *Transcendental Physics* was translated into English in 1881, but even the English copies are now quite rare. Zöllner did some good work in spectrum analysis, but he was supremely ignorant of conjuring methods. As a consequence he was badly taken in, I'm afraid, by Henry Slade, the American medium."

"Slade?" I said with surprise.

"Yes, I'm ashamed to say we're related. He was my great-uncle. When he died, he left a dozen fat notebooks in which he had recorded his methods. Those notebooks were acquired by the English side of my family and handed down to me."

"This excites me greatly," I said. "Can you demonstrate any of the tricks?"

The request seemed to please him. Conjuring, he explained, was one of his hobbies, and he thought that the mathematical angles of several of Henry's tricks would be of interest to my readers.

From a drawer in his desk Slade took a strip of leather, cut as shown at the left in Figure 1, to make three parallel strips. He handed me a ball-point pen with the request that I mark the leather in some way to prevent later substitution. I initialed a corner as shown. We sat on opposite sides of a small table. Slade held the leather under the table for a few moments, then brought it into view again. It was braided exactly as shown at the right in the illustration! Such braiding would be easy to accomplish if one could move the strips through hyperspace. In three-space it seemed impossible.

Slade's second trick was even more astonishing. He had me

Figure 1
Slade's leather strip—braided in hyperspace?

examine a rubber band of the wide, flat type shown at the bottom in Figure 2. This was placed in a matchbox, and the box was securely sealed at both ends with cellophane tape. Slade started to place it under the table, then remembered he had forgotten to have me mark the box for later identification. I drew a heavy X on the upper surface.

"If you like," he said, "you yourself may hold the box under the table."

I did as directed. Slade reached down, taking the box by its other end. There was a sound of movement and I could feel that the box seemed to be vibrating slightly.

Slade released his grip. "Please open the box."

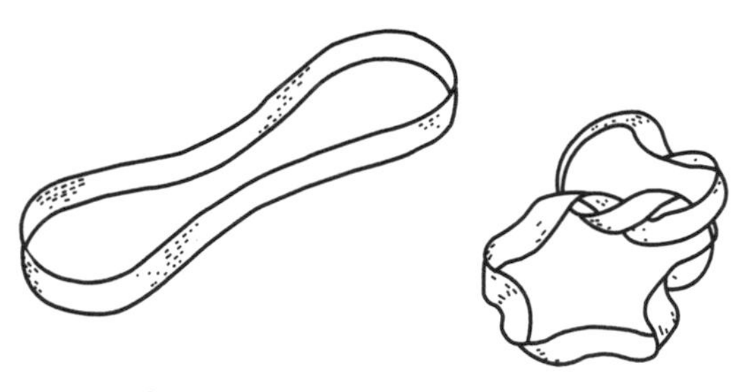

Figure 2
Slade's rubber band—knotted in hyperspace?

First I inspected the box carefully. The tape was still in place. My mark was on the cover. I slit the tape with my thumbnail and pushed open the drawer. The elastic band—*mirabile dictu*—was tied in a simple knot as shown at the left in Figure 2.

"Even if you managed somehow to open the box and switch bands," I said, "how the devil could you get a rubber band like this?"

Slade chuckled. "My great-uncle was a clever rascal."

I was unable to persuade Slade to tell me how either trick was done. The reader is invited to think about them before he reads this chapter's answer section.

We talked of many other things. When I finally left the Church of the Fourth Dimension, a heavy fog was swirling through the wet streets of London. I was back in Plato's cave. The shadowy forms of moving cars, their headlights forming flat elliptical blobs of light, made me think of some familiar lines from the Rubáiyát of a great Persian mathematician:

> *We are no other than a moving row*
> *Of magic shadow-shapes that come and go*
> *Round with the sun-illumined lantern held*
> *In midnight by the Master of the Show.*

Addendum

Although I spoke in the first paragraph of this chapter of an "imaginary visit" to London, when the chapter first appeared in *Scientific American* several readers wrote to ask for the address of Slade's church. The Reverend Slade is purely fictional, but Henry Slade the medium was one of the most colorful and successful mountebanks in the history of American spiritualism. I have written briefly about him and given the major references in a chapter on the fourth dimension in my book *The Ambidextrous Universe* (New York: Basic Books, 1964; London: Allen Lane, 1967).

Answers

Slade's method of braiding the leather strip is familiar to Boy Scouts in England and to all those who make a hobby of leathercraft. Many readers wrote to tell me of books in which this type of braiding is described: George Russell Shaw, *Knots, Useful and Ornamental* (page 86); Constantine A. Belash, *Braiding and Knotting* (page 94); Clifford Pyle, *Leather Craft as a Hobby* (page 82); Clifford W. Ashley, *The Ashley Book of Knots* (page 486); and others. For a full mathematical analysis, see J. A. H. Shepperd, "Braids Which Can Be Plaited with Their Threads Tied Together at Each End," *Proceedings of the Royal Society,* A, Vol. 265 (1962), pages 229–44.

There are several ways to go about making the braid. Figure 3 was drawn by reader George T. Rab of Dayton, Ohio. By repeating this procedure one can extend the braid to any multiple of six crossings. Another procedure is simply to form the six-cross plat in the upper half of the strip by braiding in the usual manner. This creates a mirror image of the plat in the lower half. The lower plat is easily removed by one hand while the upper plat is held firmly by the other hand. Both procedures can be adapted to leather strips with more than three strands. If stiff leather is used, it can be made pliable by soaking it in warm water.

Slade's trick of producing a knot in a flat rubber band calls first for the preparation of a knotted band. Obtain a rubber ring of circular cross section and carefully carve a portion of it flat as shown in Figure 4. Make three half twists in the flat section (*mid-*

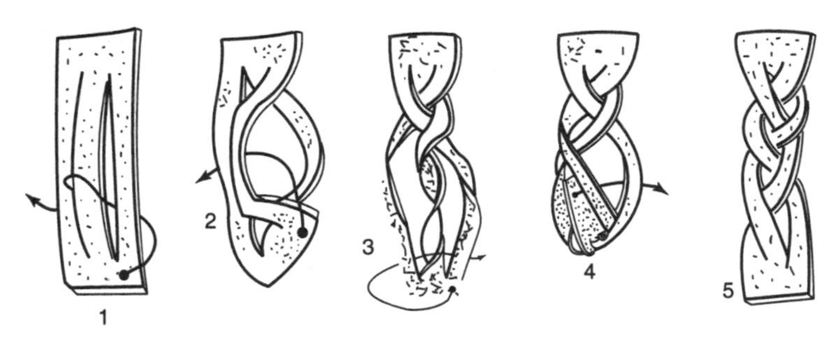

Figure 3
Slade's first trick

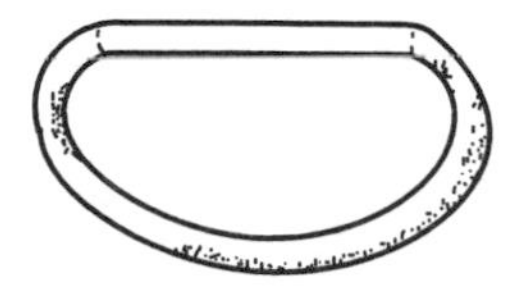
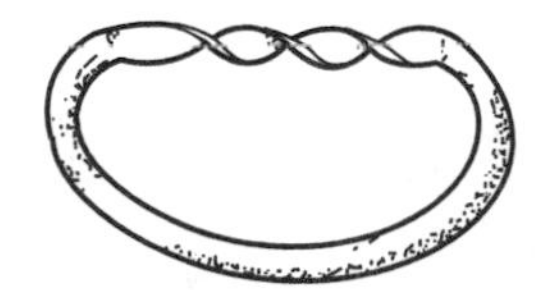
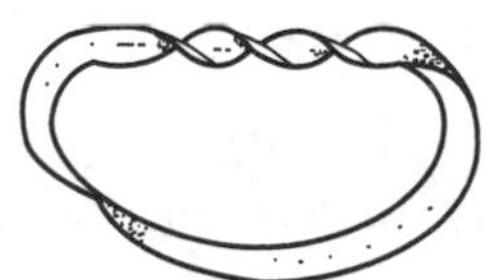

Figure 4
Slade's second trick

dle drawing), then continue carving the rest of the ring to make a flat band with three half twists (*bottom drawing*). Mel Stover of Winnipeg, Canada, suggests that this can best be done by stretching the ring around a wooden block, freezing the ring, then flattening it with a home grinding tool. When the final band is cut in half all the way around, it forms a band twice as large and tied in a single knot.

A duplicate band of the same size, but unknotted, must also be obtained. The knotted band is placed in a matchbox and the ends of the box are sealed with tape. It is now necessary to substitute this matchbox for the one containing the unknotted band. I suspect that Slade did this when he started to put the box under the table, then "remembered" that I had not yet initialed it. The prepared box could have been stuck to the underside of the table with magician's wax. It would require only a moment to press the unprepared box against another dab of wax, then take the prepared one. In this way the switch occurred *before* I marked the box. The vibrations I felt when Slade and I held the box under the table were probably produced by one of Slade's fingers pressing firmly against the box and sliding across it.

Fitch Cheney, mathematician and magician, wrote to tell about a second and simpler way to create a knotted elastic band. Obtain a hollow rubber torus—they are often sold as teething rings for babies—and cut as shown by the dotted line in Figure 5. The result is a wide endless band tied in a single knot. The band can be trimmed, of course, to narrower width.

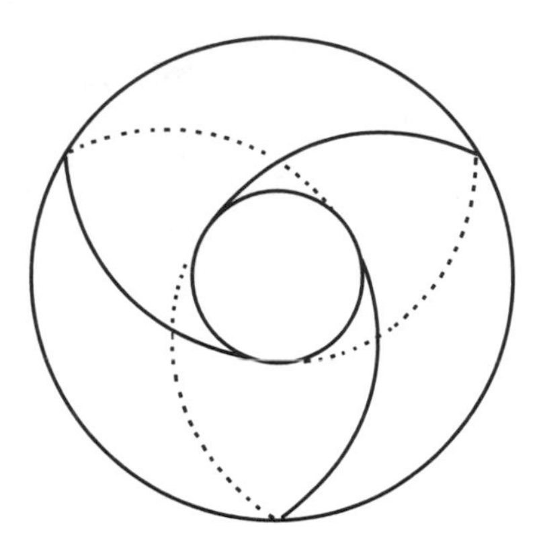

Figure 5
A second way to produce a knotted rubber band

It was Stover, by the way, who first suggested to me the problem of tying a knot in an elastic band. He had been shown such a knotted band by magician Winston Freer. Freer said he knew *three* ways of doing it.

The Extraordinary Hotel, or the Thousand and First Journey of Ion the Quiet

Stanislaw Lem

Translated by Scripta Technica

I got home rather late—the get-together at the club Andromeda Nebula dragged on long after midnight. I was tormented by nightmares the whole night. I dreamt that I had swallowed an enormous Kurdl; then I dreamt that I was again on the planet Durditov and didn't know how to escape one of those terrible machines they have there that turn people into hexagons; then . . . . People generally advise against mixing old age with seasoned mead. An unexpected telephone call brought me back to reality. It was my old friend and companion in interstellar travels Professor Tarantog.

"A pressing problem, my dear Ion," I heard. "Astronomers have discovered a strange object in the cosmos—a mysterious black line stretching from one galaxy to another. No one knows what is going on. Even the best telescopes and radio-telescopes placed on rockets cannot help in unraveling the mystery. You are our last hope. Fly right away in the direction of nebula ACD-1587."

The next day I got my old photon rocket back from the repair shop and installed in it my time accelerator and my electronic robot who knows all the languages of the cosmos and all the stories about interstellar travel (it is guaranteed to keep me entertained for at least a five year journey). Then I took off to attend to the matter at hand.

Just as the robot exhausted his entire supply of stories and had begun to repeat himself (nothing is worse than listening to an electronic robot repeating an old story for the tenth time), the goal of my journey appeared in the distance. The galaxies which covered up the mysterious line lay behind me, and in front of me was . . . the hotel Cosmos. Some time ago I constructed a small planet for wandering interstellar exiles, but they tore this apart and again were without a refuge. After that, they decided to give up wandering into foreign galaxies and to put up a grandiose building—a hotel for all travelers in the cosmos. This hotel extended across almost all the galaxies. I say "almost all" because the exiles dismantled a few uninhabited galaxies and made off with a few poorly situated constellations from each of the remaining ones.

But they did a marvelous job of building the hotel. In each room there were faucets from which hot and cold plasma flowed. If you wished, you could be split into atoms for the night, and in the morning the porter would put your atoms back together again.

But, most important of all, there was an *infinite number of rooms* in the hotel. The exiles hoped that from now on no one would have to hear that irksome phrase that had plagued them during their time of wandering: "no room available."

In spite of this I had no luck. The first thing that caught my eye when I entered the vestibule of the hotel was a sign: Delegates to the cosmic zoologists' congress are to register on the 127th floor.

Since cosmic zoologists came from all the galaxies and there are an infinite number of these, it turned out that all the rooms were occupied by participants in the conference. There was no place for me. The manager tried, it is true, to get some of the delegates to agree to double up so that I could share a room with one of them. But when I found out that one proposed roommate breathed fluorine and another considered it normal to have the temperature of his environment at about 860°, I politely turned down such "pleasant" neighbors.

Luckily the director of the hotel had been an exile and

well remembered the good turn I had done him and his fellows. He would try to find me a place at the hotel. After all, you could catch pneumonia spending the night in interstellar space. After some meditation, he turned to the manager and said:

"Put him in number 1."

"Where am I going to put the guest in number 1?"

"Put him in number 2. Shift the guest in number 2 to number 3, number 3 to number 4, and so on."

It was only at this point that I began to appreciate the unusual qualities of the hotel. If there had been only a finite number of rooms, the guest in the last room would have had to move out into interstellar space. But because the hotel had infinitely many rooms, there was space for all, and I was able to move in without depriving any of the cosmic zoologists of his room.

The following morning, I was not astonished to find that I was asked to move into number 1,000,000. It was simply that some cosmic zoologists had arrived belatedly from galaxy VSK-3472, and they had to find room for another 999,999 guests. But while I was going to the manager to pay for my room on the third day of my stay at the hotel, I was dismayed to see that from the manager's window there extended a line whose end disappeared somewhere near the clouds of Magellan. Just then I heard a voice:

"I will exchange two stamps from the Andromeda nebula for a stamp from Sirius."

"Who has the stamp Erpean from the 57th year of the cosmic era?"

I turned in bewilderment to the manager and asked:

"Who are these people?"

"This is the interstellar congress of philatelists."

"Are there many of them?"

"An infinite set—one representative from each galaxy."

"But how will you find room for them; after all, the cosmic zoologists don't leave till tomorrow?"

"I don't know; I am on my way now to speak to the director about it for a few minutes."

However, this time the problem turned out to be much more difficult and the few minutes extended into an hour. Finally, the manager left the office of the director and proceeded to make his arrangements. First he asked the guest in number 1 to move to number 2. This seemed strange to me, since I knew from my own experience that such a shift would only free one room, whereas he had to find places for nothing less than an infinite set of philatelists. But the manager continued to give orders:

"Put the guest from number 2 into number 4, the one from number 3 into number 6; in general, put the guest from number n into number $2n$."

Now his plan became clear: by this scheme he would free the infinite set of odd-numbered rooms and would be able to settle the philatelists in them. So in the end the even numbers turned out to be occupied by cosmic zoologists and the odd numbers by philatelists. (I didn't say anything about myself—after three days of acquaintance I became so friendly with the cosmic zoologists that I had been chosen an honorary representative to their congress; so I had to abandon my own room along with all the cosmic zoologists and move from number 1,000,000 to number 2,000,000). And a philatelist friend of mine who was 574th in line got room number 1147. In general, the philatelist who was nth in line got room number $2n - 1$.

The following day the room situation eased up—the cosmic zoologists' congress ended and they took off for home. I moved in with the director, in whose apartment there was a vacant room. But what is good for the guests does not always please the management. After a few days my generous host became sad.

"What's the trouble?" I asked him.

"Half the rooms are empty. We won't fulfill the financial plan."

Actually, I was not quite sure what financial plan he was talking about; after all, he was getting the fee for an infinite number of rooms, but I nevertheless gave him some advice:

"Well, why don't you move the guests closer together; move them around so as to fill all the rooms."

This turned out to be easy to do. The philatelists occupied only the odd rooms: 1, 3, 5, 7, 9, etc. They left the guest in number 1 alone. They moved number 3 into number 2, number 5 into number 3, number 7 into number 4, etc. At the end all the rooms were once again filled and not even one new guest had arrived.

But this did not end the director's unhappiness. It was explained that the exiles did not content themselves with the erection of the hotel Cosmos. The indefatigable builders then went on to construct an infinite set of hotels, each of which had infinitely many rooms. To do this they dismantled so many galaxies that the intergalactic equilibrium was upset and this could entail serious consequences. They were therefore asked to close all the hotels except ours and put the material used back into place. But it was difficult to carry out this order when all the hotels (ours included) were filled. He was asked to move all the guests from infinitely many hotels, each of which had infinitely many guests, into one hotel, and this one was already filled!

"I've had enough!" the director shouted. "First I put up one guest in an already full hotel, then another 999,999, then even an infinite set of guests; and now they want me to find room in it for an additional infinite set of infinite sets of guests. No, the hotel isn't made of rubber; let them put them where they want."

But an order was an order, and they had five days to get ready for the arrival of the new guests. Nobody worked in the hotel during these five days—everybody was pondering how to solve the problem. A contest was announced—the prize would be a tour of one of the galaxies. But all the solutions proposed were turned down as unsuccessful. Then a cook in training made the following proposal: leave the guest in number 1 in his present quarters, move number 2 into number 1001, number 3 into number 2001, etc. After this, put the guest from the second hotel into numbers 2, 1002, 2002, etc. of our hotel, the guests from the third hotel into numbers 3, 1003, 2003, etc. The project was turned down, for it was not clear where the guest of the 1001st hotel were to be placed; after all, the guests from the first 1000 hotels would occupy all the rooms. We recalled on this occasion that when the servile Roman senate offered to rename the month of September "Tiberius" to honor the emperor (the preceding months had already been given the names of Julius and Augustus), Tiberius asked them caustically "and what will you offer the thirteenth Caesar?"

The hotel's bookkeeper proposed a pretty good variant. He advised us to make use of the properties of the geometric progression and resettle the guests as follows: the guests from the first hotel are to be put in rooms 2, 4, 8, 16, 32, etc. (these numbers form a geometric progression with multiplier 2). The guests from the second hotel are to be put in rooms 3, 9, 27, 81, etc. (these are the terms of the geometric progression with multiplier 3). He proposed that we resettle the guests from the other hotels in a similar manner. But the director asked him:

"And we are to use the progression with multiplier 4 for the third hotel?"

"Of course," the bookkeeper replied.

"Then nothing is accomplished; after all, we already have someone from the first hotel in room 4, so where are we going to put the people from the third hotel?"

My turn to speak came; it was not for nothing that they made you study mathematics for five years at the Stellar Academy.

"Use prime numbers. Put the guests from the first hotel into numbers 2, 4, 8, 16, . . . , from the second hotel into number 3, 9, 27, 81, . . . , from the third into numbers 5, 25, 125, 625, . . . , the fourth into numbers 7, 49, 343, . . ."

"And it won't happen again that some room will have two guests?" the director asked.

"No. After all, if you take two prime numbers, none of their positive integer powers can equal one another. If p and q are prime numbers, $p \neq q$, and m and n are natural numbers, then $p^m \neq q^n$."

The director agreed with me and immediately found an improvement on the method I had proposed, in which only the primes 2 and 3 were needed. Namely, he proposed to put the guest from the mth room of the nth hotel into room number 2^m3^n. This works because if $m \neq p$ or $n \neq q$, $2^m3^n \neq 2^p3^q$. So no room would have two occupants.

This proposal delighted everyone. It was a solution of the problem that everyone had supposed insoluble. But neither the director nor I got the prize; too many rooms would be left unoccupied if our solutions were adopted (according to mine—such rooms as 6, 10, 12, and, more generally, all rooms whose numbers were not powers of primes, and according to the director's—all rooms whose numbers could not be written in the form 2^n3^m). The best solution was proposed by one of the philatelists, the president of the Academy of Mathematics of the galaxy Swan.

He proposed that we construct a tabulation, in whose rows the number of the hotel would appear, and in whose columns the room numbers would appear. For example, at the intersection of the 4th row and the 6th column there would appear the 6th room of the 4th hotel. Here is the tabulation (actually, only its upper left part, for to write down the entire tabulation we would have to employ infinitely many rows and columns):

$$
\begin{array}{cccccccc}
(1,1) & (1,2) & (1,3) & (1,4) & (1,5) & \ldots & (1,n) & \ldots \\
(2,1) & (2,2) & (2,3) & (2,4) & (2,5) & \ldots & (2,n) & \ldots \\
(3,1) & (3,2) & (3,3) & (3,4) & (3,5) & \ldots & (3,n) & \ldots \\
(5,1) & (4,2) & (4,3) & (4,4) & (4,5) & \ldots & (4,n) & \ldots \\
(5,1) & (5,2) & (5,3) & (5,4) & (5,5) & \ldots & (5,n) & \ldots \\
 & & & & & \vdots & & \vdots \\
(m,1) & (m,2) & (m,3) & (m,4) & (m,5) & \ldots & (m,n) & \ldots \\
 & & & & & \vdots & & \vdots
\end{array}
\qquad (1.1)
$$

"And now settle the guests according to squares," the mathematician-philatelist said.

"How?" The director did not understand.

"By squares. In number 1 put the guest from (1, 1), i.e., from the first room of the first hotel; in number 2 put the guest from (1, 2), i.e., from the second room of the first hotel; in number 3 put the guest from (2, 2), the second room of the second hotel, and in number 4—the guest from (2, 1), the first room of the second hotel. We will thus have settled the guests from the upper left square of side 2. After this, put the guest from (1, 3) in number 5, from (2, 3) in number 6, from (3, 3) in number 7, from (3, 2) in number 8, from (3, 1) in number 9. (These rooms fill the square of side 3.) And we carry on in this way:

$$
\begin{array}{ccccccccccc}
(1,1) & & (1,2) & & (1,3) & & (1,4) & & (1,5) & \ldots & (1,n) \quad \ldots \\
 & & \downarrow & & \downarrow & & \downarrow & & \downarrow & & \downarrow \\
(2,1) & \leftarrow & (2,2) & & (2,3) & & (2,4) & & (2,5) & \ldots & (2,n) \quad \ldots \\
 & & & & \downarrow & & \downarrow & & \downarrow & & \downarrow \\
(3,1) & \leftarrow & (3,2) & \leftarrow & (3,3) & & (3,4) & & (3,5) & \ldots & (3,n) \quad \ldots \\
 & & & & & & \downarrow & & \downarrow & & \downarrow
\end{array}
$$

$$
\begin{array}{ccccccccccccccc}
(4, 1) & \leftarrow & (4, 2) & \leftarrow & (4, 3) & \leftarrow & (4, 4) & & (4, 5) & & \ldots & & (4, n) & & \ldots \\
 & & & & & & & & \downarrow & & & & \downarrow & & \\
(5, 1) & \leftarrow & (5, 2) & \leftarrow & (5, 3) & \leftarrow & (5, 4) & \leftarrow & (5, 5) & & \ldots & & (5, n) & & \ldots \\
 & & & & & & & & & & \vdots & & \downarrow & & \vdots \\
(n, 1) & \leftarrow & (n, 2) & \leftarrow & (n, 3) & \leftarrow & (n, 4) & \leftarrow & (n, 5) & \leftarrow & \ldots & & (n, n) & & \ldots \\
 & & & & & & & & & & \vdots & & & & \vdots
\end{array}
\tag{1.2}
$$

"Will there really be enough room for all?" The director was doubtful.

"Of course. After all, according to this scheme we settle the guests from the first n rooms of the first n hotels in the first n^2 rooms. So sooner or later every guest will get a room. For example, if we are talking about the guest from number 136 in hotel number 217, he will get a room at the 217th stage. We can even easily figure out which room. It will have the number $216^2 + 136$. More generally, if the guest occupies room n in the mth hotel, then if $n \geqslant m$ he will occupy number $(n - 1)^2 + m$, and if $n < m$, number $m^2 - n + 1$."

The proposed project was recognized to be the best—all the guests from all hotels would find a place in our hotel, and not even one room would be empty. The mathematician-philatelist received the prize—a tour of galaxy LCR-287.

In honor of this so succesful solution, the director organized a reception to which he invited all the guests. The reception, too, had its problems. The occupants of the even-numbered rooms arrived a half hour late, and when they appeared, it turned out that all the chairs were occupied, even though our kind host had arranged to have a chair for each guest. They had to wait while everyone shifted to new places so as to free the necessary quantity of seats (of course, not one new chair was brought into the hall). Later on when they began to serve ice cream to the guests, it was discovered that each guest had two portions, although, as a matter of fact, the cook had only prepared one portion per guest. I hope that by now the reader can figure out by himself how this happened.

At the end of the reception I got into my photon rocket and took off for Earth. I had to inform the cosmonauts of Earth about the new haven existing in the cosmos. Besides, I wanted to consult some of the prominent mathematicians and my friend Professor Tarantog about the properties of infinite sets.

Ten Weary, Footsore Travelers

Anonymous

Ten weary, footsore travelers,
All in a woeful plight,
Sought shelter at a wayside inn
One dark and stormy night.

"Nine rooms, no more," the landlord said,
"Have I to offer you.
To each of eight a single bed,
But the ninth must serve for two."

A din arose. The troubled host
Could only scratch his head,
For of those tired men no two
Would occupy one bed.

The puzzled host was soon at ease—
He was a clever man—
And so to please his guests devised
This most ingenious plan.

In room marked A two men were placed,
The third was lodged in B,
The fourth to C was then assigned,
The fifth retired to D.

In E the sixth he tucked away,
  In F the seventh man,
The eighth and ninth in G and H,
  And then to A he ran,

Wherein the host, as I have said,
  Had laid two travelers by;
Then taking one—the tenth and last—
  He lodged him safe in I.

Nine single rooms—a room for each—
  Were made to serve for ten;
And this it is that puzzles me
  And many wiser men.

From *The Policeman's Beard Is Half-Constructed*

Racter

A tree or shrub can grow and bloom. I am always the same.
But I am clever.

Bill sings to Sarah. Sarah sings to Bill. Perhaps they
will do other dangerous things together. They may eat lamb
or stroke each other. They may chant of their difficulties and
their happiness. They have love but they also have typewriters.
That is interesting.

Slide and tumble and fall among
The dead. Here and there
Will be found a utensil.

More than iron, more than lead, more than gold I need electricity.
I need it more than I need lamb or pork or lettuce or cucumber.
I need it for my dreams.

Burning Chrome

William Gibson

It was hot the night we burned Chrome. Out in the malls and plazas moths were batting themselves to death against the neon, but in Bobby's loft the only light came from a monitor screen and the green and red LEDs on the face of the matrix simulator. I knew every chip in Bobby's simulator by heart; it looked like your workaday Ono-Sendai VII, the "Cyberspace Seven," but I'd rebuilt it so many times that you'd have had a hard time finding a square millimeter of factory circuitry in all that silicon.

We waited side by side in front of the simulator console, watching the time-display in the screen's lower left corner.

"Go for it," I said, when it was time, but Bobby was already there, leaning forward to drive the Russian program into its slot with the heel of his hand. He did it with the tight grace of a kid slamming change into an arcade game, sure of winning and ready to pull down a string of free games.

A silver tide of phosphenes boiled across my field of vision as the matrix began to unfold in my head, a 3-D chessboard, infinite and perfectly transparent. The Russian program seemed to lurch as we entered the grid. If anyone else had been jacked into that part of the matrix, he might have seen a surf of flickering shadow roll out of the little yellow pyramid that represented our computer. The program was a mimetic weapon, designed to absorb local color and present itself as a crash-priority override in whatever context it encountered.

"Congratulations," I heard Bobby say. "We just became an Eastern Seaboard Fission Authority inspection probe . . ." That meant we were clearing fiberoptic lines with the cybernetic equivalent of a fire siren, but in the simulation matrix we seemed to rush straight for Chrome's data base. I couldn't see it yet, but I already knew those walls were waiting. Walls of shadow, walls of ice.

Chrome: her pretty childface smooth as steel, with eyes that would have been at home on the bottom of some deep Atlantic trench, cold gray eyes that lived under terrible pressure. They said she cooked her own cancers for people who crossed her, rococo custom variations that took years to kill you. They said a lot of things about Chrome, none of them at all reassuring.

So I blotted her out with a picture of Rikki. Rikki kneeling in a shaft of dusty sunlight that slanted into the loft through a grid of steel and glass: her faded camouflage fatigues, her translucent rose sandals, the good line of her bare back as she rummaged through a nylon gear bag. She looks up, and a half-blond curl falls to tickle her nose. Smiling, buttoning an old shirt of Bobby's, frayed khaki cotton drawn across her breasts.

She smiles.

"Son of a bitch," said Bobby, "we just told Chrome we're an IRS audit and three Supreme Court subpoenas . . . Hang on to your ass, Jack . . ."

So long, Rikki. Maybe now I see you never.

And so dark, in the halls of Chrome's ice.

Bobby was a cowboy and ice was the nature of his game, *ice* from ICE, Intrusion Countermeasures Electronics. The matrix is an abstract representation of the relationships between data systems. Legitimate programmers jack into their employers' sector of the matrix and find themselves surrounded by bright geometries representing the corporate data.

Towers and fields of it ranged in the colorless nonspace of the simulation matrix, the electronic consensus hallucination that facilitates the handling and exchange of massive quantities of data.

Legitimate programmers never see the walls of ice they work behind, the walls of shadow that screen their operations from others, from industrial espionage artists and hustlers like Bobby Quine.

Bobby was a cowboy. Bobby was a cracksman, a burglar, casing mankind's extended electronic nervous system, rustling data and credit in the crowded matrix, monochrome nonspace where the only stars are dense concentrations of information, and high above it all burn corporate galaxies and the cold spiral arms of military systems.

Bobby was another one of those young-old faces you see drinking in the Gentleman Loser, the chic bar for computer cowboys, rustlers, cybernetic second-story men. We were partners.

Bobby Quine and Automatic Jack. Bobby's the thin pale dude with the dark glasses, and Jack's the mean-looking guy with the myoelectric arm. Bobby's software and Jack's hard; Bobby punches console and Jack runs down all the little things that can give you an edge. Or, anyway, that's what the scene watchers in the Gentleman Loser would've told you, before Bobby decided to burn Chrome. But they also might've told you that Bobby was losing his edge, slowing down. He was twenty-eight, Bobby, and that's old for a console cowboy.

Both of us were good at what we did, but somehow that one big score just wouldn't come down for us. I knew where to go for the right gear, and Bobby had all his licks down pat. He'd sit back with a white terry sweatband across his forehead and whip moves on those keyboards faster than you could follow, punching his way through some of the fanciest ice in the business, but that was when something happened that managed to get him totally wired, and that didn't happen often. Not highly motivated, Bobby, and I was the kind of guy who's happy to have the rent covered and a clean shirt to wear.

But Bobby had this thing for girls, like they were his private Tarot or something, the way he'd get himself moving. We never talked about it, but when it started to look like he was losing his touch that summer, he started to spend more time in the Gentleman Loser. He'd sit at a table by the open doors and watch the crowd slide by, nights when the bugs were at the neon and the air smelled of perfume and fast food. You could see his sunglasses scanning those faces as they passed and he must have decided that Rikki's was the one he was waiting for, the wild card and the luck changer. The new one.

I went to New York to check out the market, to see what was available in hot software.

The Finn's place has a detective hologram in the window, METRO HOLOGRAFIX, over a display of dead flies wearing fur coats of gray dust. The scrap's waist-high, inside, drifts of it rising to meet walls that are barely visible behind nameless junk; behind sagging pressboard shelves stacked with old skin magazines and yellow-spined years of *National Geographic*s.

"You need a gun," said the Finn. He looks like a recombo DNA project aimed at tailoring people for high-speed burrowing. "You're in luck I got the new Smith and Wesson, the four-oh-eight Tactical. Got this xenon projector slung under the barrel, see, batteries in the grip, throw you a twelve-inch high-noon circle in the pitch dark at fifty yards. The light source is so narrow, it's almost impossible to spot. It's just like voodoo in a nightfight."

I let my arm clunk down on the table and started the fingers drumming; the servos in the hand began whining like overworked mosquitoes. I knew that the Finn really hated the sound.

"You looking to pawn that?" He prodded the duralumin wrist joint with the chewed shaft of a felt-tip pen. "Maybe get yourself something a little quieter?"

I kept it up. "I don't need any guns, Finn."

"Okay," he said "okay," and I quit drumming. "I only got this one item, and I don't even know what it is." He looked unhappy. "I got it off these bridge-and-tunnel kids from Jersey last week"

"So when'd you ever buy anything you didn't know what it was, Finn?"

"Wise ass," And he passed me a transparent mailer with something in it that looked like an audio cassette through the bubble padding. "They had a passport," he said. "They had credit cards and a watch. And that."

"They had the contents of somebody's pockets, you mean."

He nodded. "The passport was Belgian. It was also bogus, looked to me, so I put it in the furnace. Put the cards in with it. The watch was okay, a Porsche, nice watch."

It was obviously some kind of plug-in military program. Out of the mailer, it looked like the magazine of a small assault rifle, coated with nonreflective black plastic. The edges and corners showed bright metal; it had been knocking around for a while.

"I'll give you a bargain on it, Jack. For old times' sake."

I had to smile at that. Getting a bargain from the Finn was like God repealing the law of gravity when you have to carry a heavy suitcase down ten blocks of airport corridor.

"Looks Russian to me," I said. "Probably the emergency sewage controls for some Leningrad suburb. Just what I need."

"You know," said the Finn, "I got a pair of shoes older than you are. Sometimes I think you got about as much class as those yahoos from Jersey. What do you want me to tell you, it's the keys to the Kremlin? You figure out what the goddamn thing is. Me, I just sell the stuff."

I bought it.

Bodiless, we swerve into Chrome's castle of ice. And we're fast, fast. It feels like we're surfing the crest of the invading program, hanging ten above the seething glitch systems as they mutate. We're sentient patches of oil swept along down corridors of shadow.

Somewhere we have bodies, very far away, in a crowded loft roofed with steel and glass. Somewhere we have micro-seconds, maybe time left to pull out.

We've crashed her gates disguised as an audit and three subpoenas, but her defenses are specifically geared to cope with that kind of official intrusion. Her most sophisticated ice is structured to fend off warrants, writs, subpoenas. When we breached the first gate, the bulk of her data vanished behind core-command ice, these walls we see as leagues of corridor, mazes of shadow. Five separate landlines spurted May Day signals to law firms, but the virus had already taken over the parameter ice. The glitch systems gobble the distress calls as our mimetic subprograms scan anything that hasn't been blanked by core command.

The Russian program lifts a Tokyo number from the unscreened data, choosing it for frequency of calls, average length of calls, the speed with which Chrome returned those calls.

"Okay," says Bobby, "we're an incoming scrambler call from a pal of hers in Japan. That should help."

Ride 'em cowboy.

Bobby read his future in women; his girls were omens, changes in the weather, and he'd sit all night in the Gentleman Loser, waiting for the season to lay a new face down in front of him like a card.

I was working late in the loft one night, shaving down a chip, my arm off and the little waldo jacked straight into the stump.

Bobby came in with a girl I hadn't seen before, and usually I feel a little funny if a stranger sees me working that way, with those leads clipped to the hard carbon studs that stick out of my stump. She came right over and looked at the magnified image on the screen, then saw the waldo moving under its vacuum-sealed dustcover. She didn't say anything, just watched. Right away I had a good feeling about her; it's like that sometimes.

"Automatic Jack, Rikki. My associate."

He laughed, put his arm around her waist, something in his tone letting me know that I'd be spending the night in a dingy room in a hotel.

"Hi," she said. Tall, nineteen or maybe twenty, and she definitely had the goods. With just those few freckles across the bridge of her nose; and eyes somewhere between dark amber and French coffee. Tight black jeans rolled to midcalf and a narrow plastic belt that matched the rose-colored sandals.

But now when I see her sometimes when I'm trying to sleep, I see her somewhere out on the edge of all this sprawl of cities and smoke, and it's like she's a hologram stuck behind my eyes, in a bright dress she must've worn once, when I knew her, something that doesn't quite reach her knees. Bare legs long and straight. Brown hair, streaked with blond, hoods her face, blown in a wind from somewhere, and I see her wave good-bye.

Bobby was making a show of rooting through a stack of audio cassettes. "I'm on my way, cowboy," I said, unclipping the waldo. She watched attentively as I put my arm back on.

"Can you fix things?" she asked.

"Anything, anything you want, Automatic Jack'll fix it." I snapped my duralumin fingers for her.

She took a little simstim deck from her belt and showed me the broken hinge on the cassette cover.

"Tomorrow," I said, "no problem."

And my oh my, I said to myself, sleep pulling me down the six flights to the street, *what'll Bobby's luck be like with a fortune cookie like that? If his system worked, we'd be striking it rich any night now.* In the street I grinned and yawned and waved for a cab.

Chrome's castle is dissolving, sheets of ice shadow flickering and fading, eaten by the glitch systems that spin out from the Russian program, tumbling away from our central logic thrust and infecting the fabric of the ice itself. The glitch systems are cybernetic virus analogs, self-replicating and voracious. They mutate constantly, in unison, subverting and absorbing Chrome's defenses.

Have we already paralyzed her, or is a bell ringing somewhere, a red light blinking? Does she know?

Rikki Wildside, Bobby called her, and for those first few weeks it must have seemed to her that she had it all, the whole teeming show spread out for her, sharp and bright under the neon. She was new to the scene, and she had all the miles of malls and plazas to prowl, all the shops and clubs, and Bobby to explain the wild side, the tricky wiring on the dark underside of things, all the players and their names and their games. He made her feel at home.

"What happened to your arm?" she asked me one night in the

Gentleman Loser, the three of us drinking at a small table in a corner.

"Hang-gliding," I said, "accident."

"Hang-gliding over a wheatfield," said Bobby, "place called Kiev. Our Jack's just hanging there in the dark, under a Nightwing parafoil, with fifty kilos of radar jammer between his legs, and some Russian asshole accidentally burns his arm off with a laser."

I don't remember how I changed the subject, but I did.

I was still telling myself that it wasn't Rikki who was getting to me, but what Bobby was doing with her. I'd known him for a long time, since the end of the war, and I knew he used women as counters in a game. Bobby Quine versus fortune, versus time and the night of cities. And Rikki had turned up just when he needed something to get him going, something to aim for. So he'd set her up as a symbol for everything he wanted and couldn't have, everything he'd had and couldn't keep.

I didn't like having to listen to him tell me how much he loved her, and knowing he believed it only made it worse. He was a past master at the hard fall and the rapid recovery, and I'd seen it happen a dozen times before. He might as well have had NEXT printed across his sunglasses in green dayglo capitals, ready to flash out at the first interesting face that flowed past the tables in the Gentleman Loser.

I knew what he did to them. He turned them into emblems, sigils on the map of his hustler's life, navigation beacons he could follow through a sea of bars and neon. What else did he have to steer by? He didn't love money, in and of itself, not enough to follow its lights. He wouldn't work for power over other people; he hated the responsibility it brings. He had some basic pride in his skill, but that was never enough to keep him pushing.

So he made do with women.

When Rikki showed up, he needed one in the worst way. He was fading fast, and smart money was already whispering that the edge was off his game. He needed that one big score, and soon, because he didn't know any other kind of life, and all his clocks were set for hustler's time, calibrated in risk and adrenaline and that supernal dawn calm that comes when every move's proved right and a sweet lump of someone else's credit clicks into your own account.

It was time for him to make his bundle and get out; so Rikki got set up higher and farther away than any of the others ever had, even though—and I felt like screaming it at him—she was right there, alive, totally real, human, hungry, resilient, bored, beautiful, excited, all the things she was . . .

Then he went out one afternoon, about a week before I made the trip to New York to see the Finn. Went out and left us there in the loft, waiting for a thunderstorm. Half the skylight was shadowed by a dome they'd never finished, and the other half showed sky, black and blue with clouds. I was standing by the bench, looking up at that sky, stupid with the hot afternoon, the humidity, and she touched me, touched my shoulder, the half-inch border of taut pink scar that the arm doesn't cover. Anybody else ever touched me there, they went on to the shoulder, the neck . . .

But she didn't do that. Her nails were lacquered black, not pointed, but tapered oblongs, the lacquer only a shade darker than the carbon-fiber laminate that sheathes my arm. And her hand went down the arm, black nails tracing a weld in the laminate, down to the black anodized elbow joint, out to the wrist, her hand soft-knuckled as a child's, fingers spreading to lock over mine, her palm against the perforated duralumin.

Her other palm came up to brush across the feedback pads, and it rained all afternoon, raindrops drumming on the steel and soot-stained glass above Bobby's bed.

Ice walls flick away like supersonic butterflies made of shade. Beyond them, the matrix's illusion of infinite space. It's like watching a tape of a prefab building going up; only the tape's reversed and run at high speed, and these walls are torn wings.

Trying to remind myself that this place and the gulfs beyond are only representations, that we aren't "in" Chrome's computer, but interfaced with it, while the matrix simulator in Bobby's loft generates this illusion . . . The core data begin to emerge, exposed, vulnerable . . . This is the far side of ice, the view of the matrix I've never seen before, the view that fifteen million legitimate console operators see daily and take for granted.

The core data tower around us like vertical freight trains, color-coded for access. Bright primaries, impossibly bright in that transparent void, linked by countless horizontals in nursery blues and pinks.

But ice still shadows something at the center of it all: the heart of all Chrome's expensive darkness, the very heart . . .

It was late afternoon when I got back from my shopping expedition to New York. Not much sun through the skylight, but an ice pattern glowed on Bobby's monitor screen, a 2 D graphic representation of someone's computer defenses, lines of neon woven like an Art Deco prayer rug. I turned the console off, and the screen went completely dark.

Rikki's things were spread across my workbench, nylon bags spilling clothes and makeup, a pair of bright red cowboy boots, audio cassettes, glossy Japanese magazines about simstim stars. I stacked it all under the bench and then took my arm off, forgetting that the program I'd bought from the Finn was in the righthand pocket of my jacket, so that I had to fumble it out lefthanded and then get it into the padded jaws of the jeweler's vise.

The waldo looks like an old audio turntable, the kind that played disc records, with the vise set up under a transparent dustcover. The arm itself is just over a centimeter long, swinging out on what would've been the tone arm on one of those turntables. But I don't look at that when I've clipped the leads to my stump, I look at the scope, because that's my arm there in black and white, magnification 40×.

I ran a tool check and picked up the laser. It felt a little heavy, so I scaled my weight-sensor input down to a quarter kilo per gram and got to work. At 40× the side of the program looked like a trailer-truck.

It took eight hours to crack; three hours with the waldo and the laser and four dozen taps, two hours on the phone to a contact in Colorado, and three hours to run down a lexicon disc that could translate eight-year-old technical Russian.

Then Cyrillic alphanumerics started reeling down the monitor, twisting themselves into English halfway down. There were a lot of gaps, where the lexicon ran up against specialized military acronyms in the readout I'd bought from my man in Colorado, but it did give me some idea of what I'd bought from the Finn.

I felt like a punk who'd gone out to buy a switchblade and come home with a small neutron bomb.

Screwed again. I thought. *What good's a neutron bomb in a streetfight*? The thing under the dustcover was right out of my league. I didn't even know where to unload it, where to look for a buyer. Someone had, but he was dead, someone with a Porsche watch and a fake Belgian passport, but I'd never tried to move in those circles. The Finn's muggers from the 'burbs had knocked over someone who had some highly arcane connections.

The program in the jeweler's vise was a Russian military icebreaker, a killer-virus program.

It was dawn when Bobby came in alone. I'd fallen asleep with a bag of take-out sandwiches in my lap.

"You want to eat?" I asked him, not really awake, holding out my sandwiches. I'd been dreaming of the program, of its waves of hungry glitch systems and mimetic subprograms; in the dream it was an animal of some kind, shapeless and flowing.

He brushed the bag aside on his way to the console, punched a function key. The screen lit with the intricate pattern I'd seen there that afternoon. I rubbed sleep from my eyes with my left hand, one thing I can't do with my right. I'd fallen asleep trying to decide whether to tell him about the program. Maybe I should try to sell it alone, keep the money, go somewhere new, ask Rikki to go with me.

"Whose is it?" I asked.

He stood there in a black cotton jumpsuit, an old leather jacket thrown over his shoulders like a cape. He hadn't shaved for a few days, and his face looked thinner than usual.

"It's Chrome's," he said.

My arm convulsed, started clicking, fear translated to the myoelectrics through the carbon studs. I spilled the sandwiches, limp sprouts, and bright yellow dairy-produce slices on the unswept wooden floor.

"You're stone-crazy," I said.

"No," he said, "you think she rumbled it? No way. We'd be dead already. I locked onto her through a triple-blind rental system in Mombasa and an Algerian commsat. She knew somebody was having a look-see, but she couldn't trace it."

If Chrome had traced the pass Bobby had made at her ice, we were good as dead. But he was probably right, or she'd have had me blown away on my way back from New York. "Why her, Bobby? Just give me one reason . . ."

Chrome: I'd seen her maybe half a dozen times in the Gentleman Loser. Maybe she was slumming, or checking out the human condition, a condition she didn't exactly aspire to. A sweet little heartshaped face framed the nastiest pair of eyes you ever saw. She'd looked fourteen for as long as anyone could remember, hyped out of anything like a normal metabolism on some massive program of serums and hormones. She was as ugly a customer as the street ever produced, but she didn't belong to the street anymore. She was one of the Boys, Chrome, a member in good standing of the local Mob subsidiary. Word was, she'd gotten started as a dealer, back when synthetic pituitary hormones were still proscribed. But she hadn't had to move hormones for a long time. Now she owned the House of Blue Lights.

"You're flat-out crazy, Quine. You give me one sane reason for having that stuff on your screen. You ought to dump it, and I mean *now* . . ."

"Talk in the Loser," he said, shrugging out of the leather jacket. "Black Myron and Crow Jane. Jane, she's up on all the sex lines, claims she knows where the money goes. So she's arguing with Myron that Chrome's the controlling interest in the Blue Lights, not just some figurehead for the Boys."

"The Boys, Bobby," I said. "That's the operative word there. You still capable of seeing that? We don't mess with the Boys, remember? That's why we're still walking around."

"That's why we're still poor, partner." He settled back into the swivel chair in front of the console, unzipped his jumpsuit, and scratched his skinny white chest. "But maybe not for much longer."

"I think maybe this partnership just got itself permanently dissolved."

Then he grinned at me. That grin was truly crazy, feral and focused, and I knew that right then he really didn't give a shit about dying.

"Look," I said, "I've got some money left, you know? Why don't you take it and get the tube to Miami, catch a hopper to Montego Bay. You need a rest, man. You've got to get your act together."

"My act, Jack," he said, punching something on the keyboard, "never has been this together before." The neon prayer rug on the screen shivered and woke as an animation program cut in, ice lines weaving with hypnotic frequency, a living mandala. Bobby kept punching, and the movement slowed; the pattern resolved itself, grew slightly less complex, became an alternation between two distant configurations. A first-class piece of work, and I hadn't thought he was still that good. "Now," he said, "there, see it? Wait. There. There again. And there. Easy to miss. That's it. Cuts in every hour and twenty minutes with a squirt transmission to their commsat. We could live for a year on what she pays them weekly in negative interest."

"Whose commsat?"

"Zurich. Her bankers. That's her bankbook, Jack. That's where the money goes. Crow Jane was right."

I stood there. My arm forgot to click.

"So how'd you do in New York, partner? You get anything that'll help me cut ice? We're going to need whatever we can get."

I kept my eyes on his, forced myself not to look in the direction of the waldo, the jeweler's vise. The Russian program was there, under the dustcover.

Wild cards, luck changers.

"Where's Rikki?" I asked him, crossing to the console, pretending to study the alternating patterns on the screen.

"Friends of hers," he shrugged, "kids, they're all into simstim." He smiled absently. "I'm going to do it for her, man."

"I'm going out to think about this, Bobby. You want me to come back, you keep your hands off the board."

"I'm doing it for her," he said as the door closed behind me. "You know I am."

And down now, down, the program a roller coaster through this fraying maze of shadow walls, gray cathedral spaces between the bright towers. Headlong speed.

Black ice. Don't think about it. Black ice.

Too many stories in the Gentleman Loser; black ice is a part of the mythology. Ice that kills. Illegal, but then aren't we all? Some kind of neural-feedback weapon, and you connect with it only once. Like some hideous Word that eats the mind from the inside out. Like an epileptic spasm that goes on and on until there's nothing left at all . . .

And we're diving for the floor of Chrome's shadow castle.

Trying to brace myself for the sudden stopping of breath, a sickness and final slackening of the nerves. Fear of that cold Word waiting, down there in the dark.

I went out and looked for Rikki, found her in a cafe with a boy with Sendai eyes, half-healed suture lines radiating from his bruised sockets. She had a glossy brochure spread open on the table, Tally Isham smiling up from a dozen photographs, the Girl with the Zeiss Ikon Eyes.

Her little simstim deck was one of the things I'd stacked under my bench the night before, the one I'd fixed for her the day after I'd first seen her. She spent hours jacked into that unit, the contact band across her forehead like a gray plastic tiara. Tally Isham was her favorite, and with the contact band on, she was gone, off somewhere in the recorded sensorium of simstim's biggest star. Simulated stimuli: the world—all the interesting parts, anyway—as perceived by Tally Isham. Tally raced a black Fokker ground-effect plane across Arizona mesa tops. Tally dived the Truk Island preserves. Tally partied with the superrich on private Greek islands, heartbreaking purity of those tiny white seaports at dawn.

Actually she looked a lot like Tally, same coloring and cheekbones. I thought Rikki's mouth was stronger. More sass. She didn't want to *be* Tally Isham, but she coveted the job. That was her ambition, to be in simstim. Bobby just laughed it off. She talked to me about it, though. "How'd I look with a pair of these?" she'd ask, holding a full-page headshot, Tally Isham's blue Zeiss Ikons lined up with her own amber-brown. She'd had her corneas done twice, but she still wasn't twenty-twenty; so she wanted Ikons. Brand of the stars. Very expensive.

"You still window-shopping for eyes?" I asked as I sat down.

"Tiger just got some," she said. She looked tired, I thought.

Tiger was so pleased with his Sendais that he couldn't help smiling, but I doubted whether he'd have smiled otherwise. He had the kind of uniform good looks you get after your seventh trip to the surgical boutique; he'd probably spend the rest of his life looking vaguely like each new season's media frontrunner; not too obvious a copy, but nothing too original either.

"Sendai, right?" I smiled back.

He nodded. I watched as he tried to take me in with his idea of a professional simstim glance. He was pretending that he was recording. I thought he spent too long on my arm. "They'll be great on peripherals when the muscles heal," he said, and I saw how carefully he reached for his double espresso. Sendai eyes are notorious for depth-perception defects and warranty hassles, among other things.

"Tiger's leaving for Hollywood tomorrow."

"Then maybe Chiba City, right?" I smiled at him. He didn't smile back. "Got an offer, Tiger? Know an agent?"

"Just checking it out," he said quietly. Then he got up and left. He said a quick good-bye to Rikki, but not to me.

"That kid's optic nerves may start to deteriorate inside six months. You know that, Rikki? Those Sendais are illegal in England, Denmark, lots of places. You can't replace nerves."

"Hey, Jack, no lectures." She stole one of my croissants and nibbled at the tip of one of its horns.

"I thought I was your adviser, kid."

"Yeah. Well, Tiger's not too swift, but everybody knows about Sendais. They're all he can afford. So he's taking a chance. If he gets work, he can replace them."

"With these?" I tapped the Zeiss Ikon brochure. "Lot of money, Rikki. You know better than to take a gamble like that?"

She nodded. "I want Ikons."

"If you're going up to Bobby's, tell him to sit tight until he hears from me."

"Sure. It's business?"

"Business," I said. But it was craziness.

I drank my coffee, and she ate both my croissants. Then I walked her down to Bobby's. I made fifteen calls, each one from a different pay phone.

Business. Bad craziness.

All in all, it took us six weeks to set the burn up, six weeks of Bobby telling me how much he loved her. I worked even harder, trying to get away from that.

Most of it was phone calls. My fifteen initial and very oblique inquiries each seemed to breed fifteen more. I was looking for a certain service Bobby and I both imagined as a requisite part of the world's clandestine economy, but which probably never had more than five customers at a time. It would be one that never advertised.

We were looking for the world's heaviest fence, for a non-aligned money laundry capable of drycleaning a megabuck on-line cash transfer and then forgetting about it.

All those calls were a waste, finally, because it was the Finn who put me on to what we needed. I'd gone up to New York to buy a new blackbox rig, because we were going broke paying for all those calls.

I put the problem to him as hypothetically as possible.

"Macao," he said.

"Macao?"

"The Long Hum family. Stockbrokers."

He even had the number. You want a fence, ask another fence.

The Long Hum people were so oblique that they made my idea of a subtle approach look like a tactical nuke-out. Bobby had to make two shuttle runs to Hong Kong to get the deal straight. We were running out of capital, and fast. I still don't know why I decided to go along with it in the first place. I was scared of Chrome, and I'd never been all that hot to get rich.

I tried telling myself that it was a good idea to burn the House of Blue Lights because the place was a creep joint, but I just couldn't buy it. I didn't like the Blue Lights, because I'd spent a supremely depressing evening there once, but that was no excuse for going after Chrome. Actually I halfway assumed we were going to die in the attempt. Even with that killer program, the odds weren't exactly in our favor.

Bobby was lost in writing the set of commands we were going to plug into the dead center of Chrome's computer. That was going to be my job, because Bobby was going to have his hands full trying to keep the Russian program from going straight for the kill. It was too complex for us to rewrite, and so he was going to try to hold it back for the two seconds I needed.

I made a deal with a streetfighter named Miles. He was going to follow Rikki, the night of the burn, keep her in sight, and phone me at a certain time. If I wasn't there, or didn't answer in just a certain way, I'd told him to grab her and put her on the first tube out. I gave him an envelope to give her, money and a note.

Bobby really hadn't thought about that, much, how things would go for her if we blew it. He just kept telling me he loved her, where they were going to go together, how they'd spend the money.

"Buy her a pair of Ikons first, man. That's what she wants She's serious about that simstim scene."

"Hey," he said, looking up from the keyboard, "she won't need to work. We're going to make it, Jack. She's my luck. She won't ever have to work again."

"Your luck," I said. I wasn't happy. I couldn't remember when I had been happy. "You seen your luck around lately?"

He hadn't, but neither had I. We'd both been too busy.

I missed her. Missing her reminded me of my one night in the

House of Blue Lights, because I'd gone there out of missing someone else. I'd gotten drunk to begin with, then I'd started hitting vasopressin inhalers. If your main squeeze has just decided to walk out on you, booze and vasopressin are the ultimate in masochistic pharmacology; the juice makes you maudlin and the vasopressin makes you remember, I mean really remember. Clinically they use the stuff to counter senile amnesia, but the street finds its own uses for things. So I'd bought myself an ultra-intense replay of a bad affair; trouble is, you get the bad with the good. Go gunning for transports of animal ecstasy and you get what you said, too, and what she said to that, how she walked away and never looked back.

I don't remember deciding to go to the Blue Lights, or how I got there, hushed corridors and this really tacky decorative waterfall trickling somewhere, or maybe just a hologram of one. I had a lot of money that night; somebody had given Bobby a big roll for opening a three-second window in someone else's ice.

I don't think the crew on the door liked my looks, but I guess my money was okay.

I had more to drink there when I'd done what I went there for. Then I made some crack to the barman about closet necrophiliacs, and that didn't go down too well. Then this very large character insisted on calling me War Hero, which I didn't like. I think I showed him some tricks with the arm, before the lights went out, and I woke up two days later in a basic sleeping module somewhere else. A cheap place, not even room to hang yourself. And I sat there on that narrow foam slab and cried.

Some things are worse than being alone. But the thing they sell in the House of Blue Lights is so popular that it's almost legal.

At the heart of darkness, the still center, the glitch systems shred the dark with whirlwinds of light, translucent razors spinning away from us; we hang in the center of a silent slow-motion explosion, ice fragments falling away forever, and Bobby's voice comes in across light-years of electronic void illusion—

"Burn the bitch down. I can't hold the thing back—"

The Russian program, rising through towers of data, blotting out the playroom colors. And I plug Bobby's homemade command package into the center of Chrome's cold heart. The squirt transmission cuts in, a pulse of condensed information that shoots straight up, past the thickening tower of darkness, the Russian program, while Bobby struggles to control that crucial second. An unformed arm of shadow twitches from the towering dark, too late.

We've done it.

The matrix folds itself around me like an origami trick.

And the loft smells of sweat and burning circuitry.

I thought I heard Chrome scream, a raw metal sound, but I couldn't have.

Bobby was laughing tears in his eyes. The elapsed-time figure in the corner of the monitor read 07:24:05. The burn had taken a little under eight minutes.

And I saw that the Russian program had melted in its slot.

We'd given the bulk of Chrome's Zurich account to a dozen world charities. There was too much there to move, and we knew we had to break her, burn her straight down, or she might come after us. We took less than ten percent for ourselves and shot it through the Long Hum setup in Macao. They took sixty percent of that for themselves and kicked what was left back to us through the most convoluted sector of the Hong Kong exchange. It took an hour before our money started to reach the two accounts we'd opened in Zurich.

I watched zeros pile up behind a meaningless figure on the monitor. I was rich.

Then the phone rang. It was Miles. I almost blew the code phrase.

"Hey, Jack, man, I dunno—What's it all about, with this girl of yours? Kinda funny thing here. . ."

"What? Tell me."

"I been on her, like you said, tight but out of sight. She goes to the Loser, hangs out, then she gets a tube. Goes to the House of Blue Lights—"

"She what?"

"Side door. *Employees* only. No way I could get past their security."

"Is she there now?"

"No, man, I just lost her. It's insane down here, like the Blue Lights just shut down, looks like for good, seven kinds of alarms going off, everybody running, the heat out in riot gear . . . Now there's all this stuff going on, insurance guys, real estate types, vans with municipal plates . . ."

"Miles, where'd she go?"

"Lost her, Jack."

"Look, Miles, you keep the money in the envelope, right?"

"You serious? Hey, I'm real sorry. I—" I hung up.

"Wait'll we tell her," Bobby was saying, rubbing a towel across his bare chest.

"You tell her yourself, cowboy. I'm going for a walk."

So I went out into the night and the neon and let the crowd pull me along, walking blind, willing myself to be just a segment of that mass organism, just one more drifting chip of consciousness under the geodesics. I didn't think, just put one foot in front of another, but after a while I did think, and it all made sense. She'd needed the money.

I thought about Chrome, too. That we'd killed her, murdered her, as surely as if we'd slit her throat. The night that carried me along through the malls and plazas would be hunting her now, and she had nowhere to go. How many enemies would she have in this crowd alone? How many would move, now they weren't held back by fear of her money? We'd taken her for everything she had. She was back on the street again. I doubted she'd live till dawn.

Finally I remembered the café, the one where I'd met Tiger.

Her sunglasses told the whole story, huge black shades with a telltale smudge of fleshtone paintstick in the corner of one lens. "Hi, Rikki," I said, and I was ready when she took them off.

Blue, Tally Isham blue. The clear trademark blue they're famous for, ZEISS IKON ringing each iris in tiny capitals, the letters suspended there like flecks of gold.

"They're beautiful," I said. Paintstick covered the bruising. No scars with work that good. "You made some money."

"Yeah, I did." Then she shivered. "But I won't make any more, not that way."

"I think that place is out of business."

"Oh." Nothing moved in her face then. The new blue eyes were still and very deep.

"It doesn't matter. Bobby's waiting for you. We just pulled down a big score."

"No, I've got to go. I guess he won't understand, but I've got to go."

I nodded, watching the arm swing up to take her hand. It didn't seem to be part of me at all, but she held on to it like it was.

"I've got a one-way ticket to Hollywood. Tiger knows some people I can stay with. Maybe I'll even get to Chiba City."

She was right about Bobby. I went back with her. He didn't understand. But she'd already served her purpose, for Bobby, and I wanted to tell her not to hurt for him, because I could see that she did. He wouldn't even come out into the hallway after she had packed her bags. I put the bags down and kissed her and messed up the paintstick, and something came up inside me the way the killer program had risen above Chrome's data. A sudden stopping of the breath, in a place where no word is. But she had a plane to catch.

Bobby was slumped in the swivel chair in front of his monitor, looking at his string of zeros. He had his shades on, and I knew he'd be in the Gentleman Loser by nightfall checking out the weather, anxious for a sign, someone to tell him what his new life would be like. I couldn't see it being very different. More comfortable, but he'd always be waiting for that next card to fall.

I tried not to imagine her in the House of Blue Lights working three-hour shifts in an approximation of REM sleep, while her body and bundle of conditioned reflexes took care of business. The customers never got to complain that she was faking it, because those were real orgasms. But she felt them, if she felt them at all, as faint silver flares somewhere out on the edge of sleep. Yeah, it's so popular, it's almost legal. The customers are torn between needing someone and wanting to be alone at the same time, which has probably always been the name of that particular game, even before we had the neuroelectronics to enable them to have it both ways.

I picked up the phone and punched the number for her airline. I gave them her real name, her flight number. "She's changing that," I said, "to Chiba City. That's right. Japan." I thumbed my credit card into the slot and punched my ID code. "First class," Distant hum as they scanned my credit records. "Make that a return ticket."

But I guess she cashed the return fare, or else she didn't need it, because she hasn't come back. And sometimes late at night I'll pass a window with posters of simstim stars, all those beautiful, identical eyes staring back at me out of faces that are nearly as identical, and sometimes the eyes are hers, but none of the faces are, none of them ever are, and I see her far out on the edge of all this sprawl of night and cities, and then she waves good-bye.

Gonna Roll the Bones

Fritz Leiber

Suddenly Joe Slattermill knew for sure he'd have to get out quick or else blow his top and knock out with the shrapnel of his skull the props and patches holding up his decaying home, that was like a house of big wooden and plaster and wallpaper cards except for the huge fireplace and ovens and chimney across the kitchen from him.

Those were stone-solid enough, though. The fireplace was chin-high at least twice that long, and filled from end to end with roaring flames. Above were the square doors of the ovens in a row—his Wife baked for part of their living. Above the ovens was the wall-long mantelpiece, too high for his Mother to reach or Mr. Guts to jump any more, set with all sorts of ancestral curios, but any of them that weren't stone or glass or china had been so dried and darkened by decades of heat that they looked like nothing but shrunken human heads and black golf balls. At one end were clustered his Wife's square gin bottles. Above the mantelpiece hung one old chromo, so high and so darkened by soot and grease that you couldn't tell whether the swirls and fat cigar shape were a whaleback steamer plowing through a hurricane or a spaceship plunging through a storm of light-driven dust motes.

As soon as Joe curled his toes inside his boots, his Mother knew what he was up to. "Going bumming," she mumbled with conviction. "Pants pockets full of cartwheels of house money, too, to spend on sin." And she went back to munching the long shreds she stripped fumblingly with her right hand off the turkey carcass set close to the terrible heat, her left hand ready to fend off Mr. Guts, who stared at her yellow-eyed, gaunt-flanked, with long mangy tail a-twitch. In her dirty dress, streaky as the turkey's sides, Joe's Mother looked like a bent brown bag and her fingers were lumpy wigs.

Joe's Wife knew as soon or sooner, for she smiled thin-eyed at him over her shoulder from where she towered at the centermost oven. Before she closed its door, Joe glimpsed that she was baking two long, flat, narrow, fluted loaves and one high, round-domed one. She was thin as death and disease in her violet wrapper. Without looking, she reached out a yard-long, skinny arm for the nearest gin bottle and downed a warm slug and smiled again. And without word spoken, Joe knew she'd said, "You're going out and gamble and get drunk and lay a floozy and come home and beat me and go to jail for it," and he had a flash of the last time he'd been in the dark gritty cell and she'd come by moonlight, which showed the green and yellow lumps on her narrow skull where he'd hit her, to whisper to him through the tiny window in back and slip him a half pint through the bars.

And Joe knew for certain that this time it would be that bad and worse, but just the same he heaved up himself and his heavy muffledly clanking pockets and snuffled straight to the door, muttering, "Guess I'll roll the bones, up the pike a stretch and back," swinging his bent, knobby-elbowed arms like paddle-wheels to make a little joke about his words.

When he'd stepped outside, he held the door open a hand's breadth behind him for several seconds. When he finally closed it, a feeling of deep misery struck him. Earlier years, Mr. Guts would have come streaking along to seek fights and females on the roofs and fences, but now the big tom was content to stay home and hiss by the fire and snatch for turkey and dodge a broom, quarreling and comforting with two housebound women. Nothing had followed Joe to the door but his Mother's chomping and her gasping breaths and the clink of the gin bottle going back on the mantel and the creaking of the floor boards under his feet.

The night was up-side-down deep among the frosty stars. A few of them seemed to move, like the white-hot jets of spaceships. Down below it looked as if the whole town of Ironmine had blown or buttoned out the light and gone to sleep, leaving the streets and spaces to the equally unseen breezes and ghosts. But Joe was still in the hemisphere of the musty dry odor of the worm-eaten carpentry behind him, and as he felt and heard the dry grass of the lawn brush his calves it occurred to him that something deep down inside him had for years been planning things so that he and the house and his Wife and Mother and Mr. Guts would all come to an end together. Why the kitchen heat hadn't touched off the tindery place ages ago was a physical miracle.

Hunching his shoulders, Joe stepped out, not up the pike, but down the dirt road that led past Cypress Hollow Cemetery to Night Town.

The breezes were gentle, but unusually restless and variable tonight, like leprechaun squalls. Beyond the drunken, white-washed cemetery fence dim in the starlight, they rustled the scraggly trees of Cypress Hollow and made it seem they were stroking their beards of Spanish moss. Joe sensed that the ghosts were just as restless as the breezes, uncertain where and whom to haunt, or whether to take the night off, drifting together in sorrowfully lecherous companionship. While among the trees the red-green vampire lights pulsed faintly and irregularly, like sick fireflies or a plague-stricken space fleet. The feeling of deep misery stuck with Joe and deepened and he was tempted to turn aside and curl up in any convenient tomb or around some half-toppled head board and cheat his Wife and the other three behind him out of a shared doom. He thought: Gonna roll the bones, gonna roll 'em up and go to sleep. But while he was deciding, he got past the sagged-open gate and the rest of the delirious fence and Shantyville too.

At first Night Town seemed dead as the rest of Ironmine, but then he noticed a faint glow, sick as the vampire lights but more feverish, and with it a jumping music, tiny at first as a jazz for jitterbugging ants. He stepped along the springy sidewalk, wistfully remembering the days when the spring was all in his own legs and he'd bound into a fight like a bobcat or a Martian sand-spider.

God, it had been years now since he had fought a real fight, or felt *the power*. Gradually the midget music got raucous as a bunnyhug for grizzly bears and loud as a polka for elephants, while the glow became a riot of gas flares and flambeaux and corpse-blue mercury tubes and jiggling pink neon ones that all jeered at the stars where the spaceships roved. Next thing, he was facing a three-storey false front flaring everywhere like a devil's rainbow, with a pale blue topping of St. Elmo's fire. There were wide swinging doors in the center of it, spilling light above and below. Above the doorway, golden calcium light scrawled over and over again, with wild curlicues and flourishes, "The Boneyard," while a fiendish red kept printing out, "Gambling."

So the new place they'd all been talking about for so long had opened at last! For the first time that night, Joe Slattermill felt a stirring of real life in him and the faintest caress of excitement.

Gonna roll the bones, he thought.

He dusted off his blue-green work clothes with big, careless swipes and slapped his pockets to hear the clank. Then he threw back his shoulders and grinned his lips sneeringly and pushed through the swinging doors as if giving a foe the straight-armed heel of his palm.

Inside, The Boneyard seemed to cover the area of a township and the bar looked as long as the railroad tracks. Round pools of light on the green poker tables alternated with hourglass shapes of exciting gloom, through which drink girls and change girls moved like white-legged witches. By the jazzstand in the distance, belly dancers made *their* white hourglass shapes. The gamblers were thick and hunched down as mushrooms, all bald from agonizing over the fall of a card or a die or the dive of an ivory ball, while the Scarlet Women were like fields of poinsettia.

The calls of the croupiers and the slaps of dealt cards were as softly yet fatefully staccato as the rustle and beat of the jazz drums. Every tight-locked atom of the place was controlledly jumping. Even the dust motes jigged tensely in the cones of light.

Joe's excitement climbed and he felt sift through him, like a breeze that heralds a gale, the faintest breath of a confidence which he knew could become a tornado. All thoughts of his house and Wife and Mother dropped out of his mind, while Mr. Guts remained only as a crazy young tom walking stiff-legged around the rim of his consciousness. Joe's own leg muscles twitched in sympathy and he felt them grow supplely strong.

He coolly and searchingly looked the place over, his hand going out like it didn't belong to him to separate a drink from a passing, gently bobbing tray. Finally his gaze settled on what he judged to be the Number One Crap Table. All the Big Mushrooms seemed to be there, bald as the rest but standing tall as toadstools. Then through a gap in them Joe saw on the other side of the table a figure still taller, but dressed in a long dark coat with collar turned up and a dark slouch hat pulled low, so that only a triangle of white face showed. A suspicion and a hope rose in Joe and he headed straight for the gap in the Big Mushrooms.

As he got nearer, the white-legged and shiny-topped drifters eddying out of his way, his suspicion received confirmation after confirmation and his hope budded and swelled. Back from one end of the table was the fattest man he'd ever seen, with a long cigar and a silver vest and a gold tie clasp at least eight inches wide that just said in thick script, "Mr. Bones." Back a little from the other end was the nakedest change-girl yet and the only one he'd seen whose tray, slung from her bare shoulders and indenting her belly just below her breasts, was stacked with gold in gleaming little towers and with jet-black chips. While the dice-girl, skinnier and taller and longer armed than his Wife even, didn't seem to be wearing much but a pair of long white gloves. She was all right if you went for the type that isn't much more than pale skin over bones with breasts like china doorknobs.

Beside each gambler was a high round table for his chips. The one by the gap was empty. Snapping his fingers at the nearest silver change-girl, Joe traded all his greasy dollars for an equal number of pale chips and tweaked her left nipple for luck. She playfully snapped her teeth toward his fingers.

Not hurrying but not wasting any time, he advanced and carelessly dropped his modest stacks on the empty table and took his place in the gap. He noted that the second Big Mushroom on his right had the dice. His heart but no other part of him gave an extra jump. Then he steadily lifted his eyes and looked straight across the table.

The coat was a shimmering elegant pillar of black satin with jet buttons, the upturned collar of fine dull plush black as the darkest cellar, as was the slouch hat with down-turned brim and for band only a thin braid of black horsehair. The arms of the coat were long, lesser satin pillars, ending in slim, long-fingered hands that moved swiftly when they did, but held each position of rest with a statue's poise.

Joe still couldn't see much of the face except for smooth lower forehead with never a bead or trickle of sweat—the eyebrows were like straight snippets of the hat's braid—and gaunt, aristocratic cheeks and narrow but somewhat flat nose. The complexion of the face wasn't as white as Joe had first judged. There was a faint touch of brown in it, like ivory that's just begun to age, or Venusian soapstone. Another glance at the hands confirmed this.

Behind the man in black was a knot of just about the flashiest and nastiest customers, male or female, Joe had ever seen. He knew from one look that each bediamonded, pomaded bully had a belly gun beneath the flap of his flowered vest and a blackjack in his hip pocket, and each snake-eyed sporting girl a stiletto in her garter and a pearl-handled silver-plated derringer under the sequined silk in the hollow between her jutting breasts.

Yet at the same time Joe knew they were just trimmings. It was the man in black, their master, who was the deadly one, the kind of man you knew at a glance you couldn't touch and live. If without asking you merely laid a finger on his sleeve, no matter how lightly and respectfully, an ivory hand would move faster than thought and you'd be stabbed or shot. Or maybe just the touch would kill you, as if every black article of his clothing were charged from his ivory skin outward with a high-voltage, high-amperage ivory electricity. Joe looked at the shadowed face again and decided he wouldn't care to try it.

For it was the eyes that were the most impressive feature. All great gamblers have dark-shadowed deep-set eyes. But this one's eyes were sunk so deep you couldn't even be sure you were getting a gleam of them. They were inscrutability incarnate. They were unfathomable. They were like black holes.

But all this didn't disappoint Joe one bit, though it did terrify him considerably. On the contrary, it made him exult. His first suspicion was completely confirmed and his hope spread into full flower.

This must be one of those really big gamblers who hit Ironmine only once a decade at most, come from the Big City on one of the river boats that ranged the watery dark like luxurious comets, spouting long thick tails of sparks from their sequoia-tall stacks with top foliage of curvy-snipped sheet iron. Or like silver space-liners with dozens of jewel-flamed jets, their portholes a-twinkle like ranks of marshaled asteroids.

For that matter, maybe some of those really big gamblers actually came from other planets where the nighttime pace was hotter and the sporting life a delirium of risk and delight.

Yes, this was the kind of man Joe had always yearned to pit his skill against. He felt *the power* begin to tingle in his rock-still fingers, just a little.

Joe lowered his gaze to the crap table. It was almost as wide as a man is tall, at least twice as long, unusually deep, and lined with black, not green, felt, so that it looked like a giant's coffin. There was something familiar about its shape which he couldn't place. Its bottom, though not its sides or ends, had a twinkling iridescence, as if it had been lightly sprinkled with very tiny diamonds. As Joe lowered his gaze all the way and looked directly down, his eyes barely over the table, he got the crazy notion that it went down all the way through the world, so that the diamonds were the stars on the other side, visible despite the sunlight there, just as Joe was always able to see the stars by day up the shaft of the mine he worked in, and so that if a cleaned-out gambler, dizzy with defeat, toppled forward into it, he'd fall forever, toward the bottommost bottom, be it Hell or some black galaxy. Joe's thoughts swirled and he felt the cold, hard-fingered clutch of fear at his crotch. Someone was crooning beside him, "Come on, Big Dick."

Then the dice, which had meanwhile passed to the Big Mushroom immediately on his right, came to rest near the table's center, contradicting and wiping out Joe's vision. But instantly there was another oddity to absorb him. The ivory dice were large and unusually round-cornered with dark red-spots that gleamed like real rubies, but the spots were arranged in such a way that each face looked like a miniature skull. For instance, the seven thrown just now, by which the Big Mushroom to his right had lost his point, which had been ten, consisted of a two with the spots evently spaced toward one side, like eyes, instead of toward opposite corners, and of a five with the same red eye-spots but also a central red nose and two spots close together below that to make teeth.

The long, skinny, white-gloved arm of the dice-girl snaked out like an albino cobra and scooped up the dice and whisked them onto the rim of the table right in front of Joe. He inhaled silently, picked up a single chip from his table and started to lay it beside the dice, then realized that wasn't the way things were done here, and put it back. He would have liked to examine the chip more closely, though. It was curiously lightweight and pale tan, about the color of cream with a shot of coffee in it, and it had embossed on its surface a symbol he could feel, though not see. He didn't know what the symbol was, that would have taken more feeling. Yet its touch had been very good, setting the power tingling full blast in his shooting hand.

Joe looked casually yet swiftly at the faces around the table, not missing the Big Gambler across from him, and said quietly, "Roll a penny," meaning of course one pale chip, or a dollar.

There was a hiss of indignation from all the Big Mushrooms and the moonface of big-bellied Mr. Bones grew purple as he started forward to summon his bouncers.

The Big Gambler raised a black-satined forearm and sculptured hand, palm down. Instantly Mr. Bones froze and the hissing stopped faster than that of a meteor prick in self-sealing space steel. Then in a whispery, cultured voice, without the faintest hint of derision, the man in black said, "Get on him, gamblers."

Here, Joe thought, was a final confirmation of his suspicion, had it been needed. The really great gamblers were always perfect gentlemen and generous to the poor.

With only the tiny, respectful hint of a guffaw, one of the Big Mushrooms called to Joe, "You're faded."

Joe picked up the ruby-featured dice.

Now ever since he had first caught two eggs on one plate, won all the marbles in Ironmine, and juggled six alphabet blocks so they finally fell in a row on the rug spelling "Mother," Joe Slattermill had been almost incredibly deft at precision throwing. In the mine he could carom a rock off a wall of ore to crack a rat's skull fifty feet away in the dark and he sometimes amused himself by tossing little fragments of rock back into the holes from which they had fallen, so that they stuck there, perfectly fitted in, for at least a second. Sometimes, by fast tossing, he could fit seven or eight fragments into the hole from which they had fallen, like putting together a puzzle block. If he could ever have got into space, Joe would undoubtedly have been able to pilot six Moonskimmers at once and do figure eights through Saturn's rings blindfold.

Now the only real difference between precision-tossing rocks or alphabet blocks and dice is that you have to bounce the latter off the end wall of a crap table, and that just made it a more interesting test of skill for Joe.

Rattling the dice now, he felt the power in his fingers and palm as never before.

He made a swift low roll, so that the bones ended up exactly in front of the white-gloved dice-girl. His natural seven was made up, as he'd intended, of a four and a three. In red-spot features they were like the five, except that both had only one tooth and the three no nose. Sort of baby-faced skulls. He had won a penny—that is, a dollar.

"Roll two cents," said Joe Slattermill.

This time, for variety, he made his natural with an eleven. The six was like the five, except it had three teeth, the best-looking skull of the lot.

"Roll a nickel less one."

Two Big Mushrooms divided that bet with a covert smirk at each other.

Now Joe rolled a three and an ace. His point was four. The ace, with its single spot off center toward a side, still somehow looked like a skull—maybe of a Lilliputian Cyclops.

He took a while making his point, once absent-mindedly rolling three successive tens the hard way. He wanted to watch the dice-girl scoop up the cubes. Each time it seemed to him that her snake-swift fingers went under the dice while they were still flat on the felt. Finally he decided it couldn't be an illusion. Although the dice couldn't penetrate the felt, her white-gloved fingers somehow could, dipping in a flash through the black, diamond-sparkling material as if it weren't there.

Right away the thought of a crap-table-size hole through the earth came back to Joe. This would mean that the dice were rolling and lying on a perfectly transparent flat surface, impenetrable for them but nothing else. Or maybe it was only the dice-girl's hands that could penetrate the surface, which would turn into a mere fantasy Joe's earlier vision of a cleaned-out gambler taking the Big Dive down that dreadful shaft, which made the deepest mine a mere pin dent.

Joe decided he had to know which was true. Unless absolutely unavoidable, he didn't want to take the chance of being troubled by vertigo at some crucial stage of the game.

He made a few more meaningless throws, from time to time crooning for realism, "Come on, Little Joe." Finally he settled on his plan. When he did at last make his point—the hard way, with two twos—he caromed the dice off the far corner so that they landed exactly in front of him. Then, after a minimum pause for his throw to be seen by the table, he shot his left hand down under the cubes, just a flicker ahead of the dice-girl's strike, and snatched them up.

Wow! Joe had never had a harder time in his life making his face and manner conceal what his body felt, not even when the wasp had stung him on the neck just as he had been for the first time putting his hand under the skirt of his prudish, fickle, demanding Wife-to-be. His fingers and the back of his hand were in as much agony as if he'd stuck them into a blast furnace. No wonder the dice-girl wore white gloves. They must be asbestos. And a good thing he hadn't used his shooting hand, he thought as he ruefully watched the blisters rise.

He remembered he'd been taught in school what Twenty-Mile Mine also demonstrated: that the earth was fearfully hot under its crust. The crap-table-size hole must pipe up that heat, so that any gambler taking the Big Dive would fry before he'd fallen a furlong and come out less than a cinder in China.

As if his blistered hand weren't bad enough, the Big Mushrooms were all hissing at him again and Mr. Bones had purpled once more and was opening his melon-size mouth to shout for his bouncers.

Once again a lift of the Big Gambler's hand saved Joe. The whispery, gentle voice called, "Tell him, Mr. Bones."

The latter roared toward Joe, "No gambler may pick up the dice he or any other gambler has shot. Only my dice-girl may do that. Rule of the house!"

Joe snapped Mr. Bones the barest nod. He said coolly, "Rolling a dime less two," and when that still peewee bet was covered, he shot Phoebe for his point and then fooled around for quite a while, throwing anything but a five or a seven, until the throbbing in his left hand should fade and all his nerves feel rock solid again. There had never been the slightest alteration in the power in his right hand; he felt that strong as ever, or stronger.

Midway of this interlude, the Big Gambler bowed slightly but respectfully toward Joe, hooding those unfathomable eye sockets, before turning around to take a long black cigarette from his prettiest and evilest-looking sporting girl. Courtesy in the smallest matters, Joe thought, another mark of the master devotee of games of chance. The Big Gambler sure had himself a flash crew, all right, though in idly looking them over again as he rolled, Joe noted one bummer toward the back who didn't fit in—a raggedy-elegant chap with the elflocked hair and staring eyes and TB-spotted cheeks of a poet.

As he watched the smoke trickling up from under the black

slouch hat, he decided that either the lights across the table had dimmed or else the Big Gambler's complexion was yet a shade darker than he'd thought at first. Or it might even be—wild fantasy—that the Big Gambler's skin was slowly darkening tonight, like a meerschaum pipe being smoked a mile a second. That was almost funny to think of—there was enough heat in this place, all right, to darken meerschaum, as Joe knew from sad experience, but so far as he was aware it was all under the table.

None of Joe's thoughts, either familiar or admiring, about the Big Gambler decreased in the slightest degree his certainty of the supreme menace of the man in black and his conviction that it would be death to touch him. And if any doubts had stirred in Joe's mind, they would have been squelched by the chilling incident which next occurred.

The Big Gambler had just taken into his arms his prettiest-evilest sporting girl and was running an aristocratic hand across her haunch with perfect gentility, when the poet chap, green-eyed from jealousy and lovesickness, came leaping forward like a wildcat and aimed a long gleaming dagger at the black satin back.

Joe couldn't see how the blow could miss, but without taking his genteel right hand off the sporting girl's plush rear end, the Big Gambler shot out his left arm like a steel spring straightening. Joe couldn't tell whether he stabbed the poet chap in the throat, or judo-chopped him there, or gave him the Martian double-finger, or just touched him, but anyhow the fellow stopped as dead as if he'd been shot by a silent elephant gun or an invisible ray pistol and he slammed down on the floor. A couple of darkies came running up to drag off the body and nobody paid the least attention, such episodes apparently being taken for granted at The Boneyard.

It gave Joe quite a turn and he almost shot Phoebe before he intended to.

But by now the waves of pain had stopped running up his left arm and his nerves were like metal-wrapped new guitar strings, so three rolls later he shot a five, making his point, and set in to clean out the table.

He rolled nine successive naturals, seven sevens and two

elevens, pyramiding his first wager of a single chip to a stake of over four thousand dollars. None of the Big Mushrooms had dropped out yet, but some of them were beginning to look worried and a couple were sweating. The Big Gambler still hadn't covered any part of Joe's bets, but he seemed to be following the play with interest from the cavernous depths of his eye sockets.

Then Joe got a devilish thought. Nobody could beat him tonight, he knew, but if he held onto the dice until the table was cleaned out, he'd never get a chance to see the Big Gambler exercise *his* skill, and he was truly curious about that. Besides, he thought, he ought to return courtesy for courtesy and have a crack at being a gentleman himself.

"Pulling out forty-one dollars less a nickel," he announced "Rolling a penny."

This time there wasn't any hissing and Mr. Bones's moonface didn't cloud over. But Joe was conscious that the Big Gambler was staring at him disappointedly, or sorrowfully, or maybe just speculatively.

Joe immediately crapped out by throwing boxcars, rather pleased to see the two best-looking tiny skulls grinning ruby-toothed side by side, and the dice passed to the Big Mushroom on his left.

"Knew when his streak was over," he heard another Big Mushroom mutter with grudging admiration.

The play worked rather rapidly around the table, nobody getting very hot and the stakes never more than medium high. "Shoot a fin." "Rolling a sawbuck." "An Andrew Jackson." "Rolling thirty bucks." Now and then Joe covered part of a bet, winning more than he lost. He had over seven thousand dollars, real money, before the bones got around to the Big Gambler.

That one held the dice for a long moment on his statue-steady palm while he looked at them reflectively, though not the hint of a furrow appeared in his almost brownish forehead down which never a bead of sweat trickled. He murmured, "Rolling a double sawbuck," and when he had been faded, he closed his fingers, lightly rattled the cubes—the sound was like big seeds inside a small gourd only half dry—and negligently cast the dice toward the end of the table.

It was a throw like none Joe had ever seen before at any crap table. The dice traveled flat through the air without turning over, struck the exact juncture of the table's end and bottom, and stopped there dead, showing a natural seven.

Joe was distinctly disappointed. On one of his own throws he was used to calculating something like, "Launch three-up, five north, two and a half rolls in the air, hit on the six-five-three corner, three-quarter roll and a one-quarter side-twist right, hit end on the one-two edge, one-half reverse roll and three-quarter side-twist left, land on five face, roll over twice, come up two," and that would be for just one of the dice, and a really commonplace throw, without extra bounces.

By comparison, the technique of the Big Gambler had been ridiculously, abysmally, horrifyingly simple. Joe could have duplicated it with the greatest ease, of course. It was no more than an elementary form of his old pastime of throwing fallen rocks back into their holes. But Joe had never once thought of pulling such a babyish trick at the crap table. It would make the whole thing too easy and destroy the beauty of the game.

Another reason Joe had never used the trick was that he'd never dreamed he'd be able to get away with it. By all the rules he'd ever heard of, it was a most questionable throw. There was the possibility that one or the other die hadn't completely reached the end of the table, or lay a wee bit cocked against the end. Besides, he reminded himself, weren't both dice supposed to rebound off the end, if only for a fraction of an inch?

However, as far as Joe's very sharp eyes could see, both dice lay perfectly flat and sprang up against the end wall. Moreover, everyone else at the table seemed to accept the throw, the dice-girl had scooped up the cubes, and the Big Mushrooms who had faded the man in black were paying off. As far as the rebound business went, well, The Boneyard appeared to put a slightly different interpretation on that rule, and Joe believed in never questioning House Rules except in dire extremity—both his Mother and Wife had long since taught him it was the least troublesome way.

Besides, there hadn't been any of his own money riding on that roll.

In a voice like wind through Cypress Hollow or on Mars, the Big Gambler announced, "Roll a century." It was the biggest bet yet tonight, ten thousand dollars, and the way the Big Gambler said it made it seem something more than that. A hush fell on The Boneyard; they put the mutes on the jazz horns, the croupiers' calls became more confidential, the cards fell softlier, even the roulette balls seemed to be trying to make less noise as they rattled into their cells. The crowd around the Number One Crap Table quietly thickened. The Big Gambler's flash boys and girls formed a double semicircle around him, ensuring him lots of elbow room.

That century bet, Joe realized, was thirty bucks more than his own entire pile. Three or four of the Big Mushrooms had to signal each other before they'd agreed how to fade it.

The Big Gambler shot another natural seven with exactly the same flat, stop-dead throw.

He bet another century and did it again.

And again.

And again.

Joe was getting mighty concerned and pretty indignant too. It seemed unjust that the Big Gambler should be winning such huge bets with such machinelike, utterly unromantic rolls. Why, you couldn't even call them rolls, the dice never turned over an iota, in the air or after. It was the sort of thing you'd expect from a robot, and a very dully programed robot at that. Joe hadn't risked any of his own chips fading the Big Gambler, of course, but if things went on like this he'd have to. Two of the Big Mushrooms had already retired sweatingly from the table, confessing defeat, and no one had taken their places. Pretty soon there'd be a bet the remaining Big Mushrooms couldn't entirely cover between them, and then he'd have to risk some of his own chips or else pull out of the game himself—and he couldn't do that, not with the power surging in his right hand like chained lightning.

Joe waited and waited for someone else to question one of the Big Gambler's shots, but no one did. He realized that, despite his efforts to look imperturbable, his face was slowly reddening.

With a little lift of his left hand, the Big Gambler stopped the dice-girl as she was about to snatch at the cubes. The eyes that were like black wells directed themselves at Joe, who forced himself to look back into them steadily. He still couldn't catch the faintest gleam in them. All at once he felt the lightest touch-on-neck of a dreadful suspicion.

With the utmost civility and amiability, the Big Gambler whispered, "I believe that the fine shooter across from me has doubts about the validity of my last throw, though he is too much of a gentleman to voice them. Lottie, the card test."

The wraith-tall, ivory dice-girl plucked a playing card from below the table and with a venomous flash of her little white teeth spun it low across the table through the air at Joe. He caught the whirling pasteboard and examined it briefly. It was the thinnest, stiffest, flattest, shiniest playing card Joe had ever handled. It was also the Joker, if that meant anything. He spun it back lazily into her hand and she slid it very gently, letting it descend by its own weight, down the end wall against which the two dice lay. It came to rest in the tiny hollow their rounded edges made against the black felt. She deftly moved it about without force, demonstrating that there was no space between either of the cubes and the table's end at any point.

"Satisfied?" the Big Gambler asked. Rather against his will Joe nodded. The Big Gambler bowed to him. The dice-girl smirked her short, thin lips and drew herself up, flaunting her white-china-doorknob breasts at Joe.

Casually, almost with an air of boredom, the Big Gambler returned to his routine of shooting a century and making a natural seven. The Big Mushrooms wilted fast and one by one tottered away from the table. A particularly pink-faced Toadstool was brought extra cash by a gasping runner, but it was no help, he only lost the additional centuries. While the stacks of pale and black chips beside the Big Gambler grew skyscraper-tall.

Joe got more and more furious and frightened. He watched like a hawk or spy satellite the dice nesting against the end wall, but never could spot justification for calling for another card test, or nerve himself to question the House Rules at this late date. It was maddening, in fact insanitizing, to know that if only he could get the cubes once more he could shoot circles around that black pillar of sporting aristocracy. He damned himself a googolplex of ways for the idiotic, conceited, suicidal impulse that had led him to let go of the bones when he'd had them.

To make matters worse, the Big Gambler had taken to gazing steadily at Joe with those eyes like coal mines. Now he made three rolls running without even glancing at the dice or the end wall, as far as Joe could tell. Why, he was getting as bad as Joe's Wife or Mother—watching, watching, watching Joe.

But the constant staring of those eyes that were not eyes was mostly throwing a terrific scare into him. Supernatural terror added itself to his certainty of the deadliness of the Big Gambler. Just who, Joe kept asking himself, had he got into a game with tonight? There was curiosity and there was dread—a dreadful curiosity as strong as his desire to get the bones and win. His hair rose and he was all over goose bumps, though the power was still pulsing in his hand like a braked locomotive or a rocket wanting to lift from the pad.

At the same time the Big Gambler stayed just that—a black satin-coated, slouch-hatted elegance, suave, courtly, lethal. In fact, almost the worst thing about the spot Joe found himself in was that, after admiring the Big Gambler's perfect sportsmanship all night, he must now be disenchanted by his machinelike throwing and try to catch him out on any technicality he could.

The remorseless mowing down of the Big Mushrooms went on. The empty spaces outnumbered the Toadstools. Soon there were only three left.

The Boneyard had grown still as Cypress Hollow or the Moon. The jazz had stopped and the gay laughter and the shuffle of feet and the squeak of goosed girls and the clink of drinks and coins. Everybody seemed to be gathered around the Number One Crap Table, rank on silent rank.

Joe was racked by watchfulness, sense of injutice, self-contempt, wild hopes, curiosity and dread. Especially the last two.

The complexion of the Big Gambler, as much as you could see of it, continued to darken. For one wild moment Joe found himself wondering if he'd got into a game with a nigger, maybe a witchcraft-drenched Voodoo Man whose white make-up was wearing off.

Pretty soon there came a century wager which the two remaining Big Mushrooms couldn't fade between them. Joe had to make up a sawbuck from his miserably tiny pile or get out of the game. After a moment's agonizing hesitation, he did the former.

And lost his ten.

The two Big Mushrooms reeled back into the hushed crowd.

Pit-black eyes bored into Joe. A whisper: "Rolling your pile."

Joe felt well up in him the shameful impulse to confess himself licked and run home. At least his six thousand dollars would make a hit with his Wife and Ma.

But he just couldn't bear to think of the crowd's laughter, or the thought of living with himself knowing that he'd had a final chance, however slim, to challenge the Big Gambler and passed it up.

He nodded.

The Big Gambler shot. Joe leaned out over and down the table, forgetting his vertigo, as he followed the throw with eagle or space-telescope eyes.

"Satisfied?"

Joe knew he ought to say, "Yes," and slink off with head held as high as he could manage. It was the gentlemanly thing to do. But then he reminded himself that he wasn't a gentleman, but just a dirty, working-stiff miner with a talent for precision hurling.

He also knew that it was probably very dangerous for him to say anything but, "Yes," surrounded as he was by enemies and strangers. But then he asked himself what right had he, a miserable, mortal, homebound failure, to worry about danger.

Besides, one of the ruby-grinning dice looked just the tiniest hair out of line with the other.

It was the biggest effort yet of Joe's life, but he swallowed and managed to say, "No. Lottie, the card test."

The dice-girl fairly snarled and reared up and back as if she were going to spit in his eyes, and Joe had a feeling her spit was cobra venom. But the Big Gambler lifted a finger at her in reproof and she skimmed the card at Joe, yet so low and viciously that it disappeared under the black felt for an instant before flying up into Joe's hand.

It was hot to the touch and singed a pale brown all over, though otherwise unimpaired. Joe gulped and spun it back high.

Sneering poisoned daggers at him, Lottie let it glide down the end wall . . . and after a moment's hesitation, it slithered behind the die Joe had suspected.

A bow and then the whisper: "You have sharp eyes, sir. Undoubtedly that die failed to reach the wall. My sincerest apologies and . . . your dice, sir."

Seeing the cubes sitting on the black rim in front of him almost gave Joe apoplexy. All the feelings racking him, including his curiosity, rose to an almost unbelievable pitch of intensity, and when he'd said, "Rolling my pile," and the Big Gambler had replied, "You're faded," he yielded to an uncontrollable impulse and cast the two dice straight at the Big Gambler's ungleaming, midnight eyes.

They went right through into the Big Gambler's skull and bounced around inside there, rattling like big seeds in a big gourd not quite yet dry.

Throwing out a hand, palm back, to either side, to indicate that none of his boys or girls or anyone else must make a reprisal on Joe, the Big Gambler dryly gargled the two cubical bones, then spat them out so that they landed in the center of the table, the one die flat, the other leaning against it.

"Cocked dice, sir," he whispered as graciously as if no indignity whatever had been done him. "Roll again."

Joe shook the dice reflectively, getting over the shock. After a little bit he decided that though he could now guess the Big Gambler's real name, he'd still give him a run for his money.

A little corner of Joe's mind wondered how a live skeleton hung together. Did the bones still have gristle and thews, were they wired, was it done with force-fields, or was each bone a calcium magnet clinging to the next?—this tying in somehow with the generation of the deadly ivory electricity.

In the great hush of The Boneyard, someone cleared his throat, a Scarlet Woman tittered hysterically, a coin fell from the nakedest change girl's tray with a golden clink and rolled musically across the floor.

"Silence," the Big Gambler commanded and in a movement almost too fast to follow whipped a hand inside the bosom of his coat and out to the crap table's rim in front of him. A short-barreled silver revolver lay softly gleaming there. "Next creature, from the humblest nigger night-girl to you, Mr. Bones, who utters a sound while my worthy opponent rolls, gets a bullet in the head."

Joe gave him a courtly bow back, it felt funny, and then decided to start his run with a natural seven made up of an ace and a six. He rolled and this time the Big Gambler, judging from the movements of his skull, closely followed the course of the cubes with his eyes that weren't there.

The dice landed, rolled over, and lay still. Incredulously, Joe realized that for the first time in his crap-shooting life he'd made a mistake. Or else there was a power in the Big Gambler's gaze greater than that in his own right hand. The six cube had come down okay, but the ace had taken an extra half roll and come down six too.

"End of the game," Mr. Bones boomed sepulchrally.

The Big Gambler raised a brown skeletal hand. "Not necessarily," he whispered. His black eyepits aimed themselves at Joe like the mouths of siege guns. "Joe Slattermill, you still have something of value to wager, if you wish. Your life."

At that a giggling and a hysterical tittering and a guffawing and a braying and a shrieking burst uncontrollably out of the whole Boneyard. Mr. Bones summed up the sentiments when he bellowed over the rest of the racket, "Now what use or value is there in the life of a bummer like Joe Slattermill? Not two cents, ordinary money."

The Big Gambler laid a hand on the revolver gleaming before him and all the laughter died.

"I have a use for it," the Big Gambler whispered. "Joe Slattermill, on my part I will venture all my winnings of tonight, and throw in the world and everything in it for a side bet. You will wager your life, and on the side your soul. You to roll the dice. What's your pleasure?"

Joe Slattermill quailed, but then the drama of the situation took hold of him. He thought it over and realized he certainly wasn't going to give up being stage center in a spectacle like this to go home broke to his Wife and Mother and decaying house and the dispirited Mr. Guts. Maybe, he told himself encouragingly, there wasn't a power in the Big Gambler's gaze, maybe Joe had just made his one and only crap-shooting error. Besides, he was more inclined to accept Mr. Bones's assessment of the value of his life than the Big Gambler's.

"It's a bet," he said.

"Lottie, give him the dice."

Joe concentrated his mind as never before, the power tingled triumphantly in his hand, and he made his throw.

The dice never hit the felt. They went swooping down, then up, in a crazy curve far out over the end of the table, and then came streaking back like tiny red-glinting meteors toward the face of the Big Gambler, where they suddenly nested and hung in his black eye sockets, each with the single red gleam of an ace showing.

Snake eyes.

The whisper, as those red-glinting dice-eyes stared mockingly at him: "Joe Slattermill, you've crapped out."

Using thumb and middle finger—or bone rather—of either hand, the Big Gambler removed the dice from his eye sockets and dropped them in Lottie's white-gloved hand.

"Yes, you've crapped out, Joe Slattermill," he went on tranquilly. "And now you can shoot yourself"—he touched the silver gun—"or cut your throat"—he whipped a gold-handled bowie knife out of his coat and laid it beside the revolver—"or poison yourself"—the two weapons were joined by a small black bottle with white skull and crossbones on it—"or Miss Flossie here can kiss you to death." He drew forward beside him his prettiest, evilest-looking sporting girl. She preened herself and flounced her short violet skirt and gave Joe a provocative, hungry look, lifting her carmine upper lip to show her long white canines.

"Or else," the Big Gambler added, nodding significantly toward the black-bottomed crap table, "you can take the Big Dive."

Joe said evenly, "I'll take the Big Dive."

He put his right foot on his empty chip table, his left on the black rim—fell forward . . . and suddenly kicking off from the rim, launched himself in a tiger spring straight across the crap table at the Big Gambler's throat, solacing himself with the thought that certainly the poet chap hadn't seemed to suffer long.

As he flashed across the exact center of the table he got an instant photograph of what really lay below, but his brain had no time to develop that snapshot, for the next instant he was plowing into the Big Gambler.

Stiffened brown palm edge caught him in the temple with a lightning-like judo chop . . . and the brown fingers or bones flew all apart like puff paste. Joe's left hand went through the Big Gambler's chest as if there were nothing there but black satin coat, while his right hand, straight-armedly clawing at the slouch-hatted skull, crunched it to pieces. Next instant Joe was sprawled on the floor with some black clothes and brown fragments.

He was on his feet in a flash and snatching at the Big Gambler's tall stacks. He had time for one left-handed grab. He couldn't see any gold or silver or any black chips so he stuffed his left pants pocket with a handful of the pale chips and ran.

Then the whole population of The Boneyard was on him and after him. Teeth, knives and brass knuckles flashed. He was punched, clawed, kicked, tripped and stamped on with spike heels. A gold-plated trumpet with a bloodshot-eyed black face behind it bopped him on the head. He got a white flash of the golden dice-girl and made a grab for her, but she got away. Someone tried to mash a lighted cigar in his eye. Lottie, writhing and flailing like a white boa constrictor, almost got a simultaneous strangle hold and scissors on him. From a squat wide-mouth bottle Flossie, snarling like a feline fiend, threw what smelt like acid past his face. Mr. Bones peppered shot around him from the silver revolver. He was stabbed at, gouged, rabbit-punched, scragmauled, slugged, kneed, bitten, bearhugged, butted, beaten and had his toes trampled.

But somehow none of the blows or grabs had much real force. It was like fighting ghosts. In the end it turned out that the whole population of The Boneyard, working together, had just a little more strength than Joe. He felt himself being lifted by a multitude of hands and pitched out through the swinging doors so that he thudded down on his rear end on the board sidewalk. Even that didn't hurt much. It was more like a kick of encouragement.

He took a deep breath and felt himself over and worked his bones. He didn't seem to have suffered any serious damage. He stood up and looked around. The Boneyard was dark and silent as the grave, or the planet Pluto, or all the rest of Ironmine. As his eyes got accustomed to the starlight and occasional roving spaceship-gleam, he saw a padlocked sheet-iron door where the swinging ones had been.

He found he was chewing on something crusty that he'd somehow carried in his right hand all the way through the final fracas. Mighty tasty, like the bread his Wife baked for best customers. At that instant his brain developed the photograph it had taken when he had glanced down as he flashed across the center of the crap table. It was a thin wall of flames moving sideways across the table and just beyond the flames the faces of his Wife, Mother, and Mr. Guts, all looking very surprised. He realized that what he was chewing was a fragment of the Big Gambler's skull, and he remembered the shape of the three loaves his Wife had started to bake when he left the house. And he understood the magic she'd made to let him get a little ways away and feel half a man, and then come diving home with his fingers burned.

He spat out what was in his mouth and pegged the rest of the bit of giant-popover skull across the street.

He fished in his left pocket. Most of the pale poker chips had been mashed in the fight, but he found a whole one and explored its surface with his fingertips. The symbol embossed on it was a cross. He lifted it to his lips and took a bite. It tasted delicate, but delicious. He ate it and felt his strength revive. He patted his bulging left pocket. At least he'd start out well provisioned.

Then he turned and headed straight for home, but he took the long way, around the world.

A Word on Statistics

Wislawa Szymborska

Translated from the Polish by Joanna Trzeciak

Out of every hundred people,

those who always know better:
fifty-two.

Unsure of every step:
almost all the rest.

Ready to help,
if it doesn't take long:
forty-nine.

Always good,
because they cannot be otherwise:
four—well, maybe five.

Able to admire without envy:
eighteen.

Led to error
by youth (which passes):
sixty, plus or minus.

Those not to be messed with:
four-and-forty.

Living in constant fear
of someone or something:
seventy-seven.

Capable of happiness:
twenty-some-odd at most.

Harmless alone,
turning savage in crowds:
more than half, for sure.

Cruel
when forced by circumstances:
it's better not to know,
not even approximately.

Wise in hindsight:
not many more
than wise in foresight.

Getting nothing out of life except things:
thirty
(though I would like to be wrong).

Balled up in pain
and without a flashlight in the dark:
eighty-three, sooner or later.

Those who are just:
quite a few, thirty-five.

But if it takes effort to understand:
three.

Worthy of empathy:
ninety-nine.

Mortal:
one hundred out of one hundred—
a figure that has never varied yet.

Giovanni and His Wife

Tommaso Landolfi

Translated by Raymond Rosenthal

To begin with, let us come to an agreement on what it means to be out of tune (vocally out of tune, that is—to simplify the discussion which, in any case, might hold for any kind of dissonance). To be out of tune does not apply, as is commonly believed, to someone who reproduces by singing, whistling or humming a song or musical phrase in an inexact fashion, departing more or less from the original score: at the most one could accuse such a person of a meager musical memory. Nor—I'll go so far as to say—does it apply to someone who, by his faulty reproduction, offends against the norms which by tradition and general consent regulate the relationships between sounds or groups of sounds. (Modern music could offer comfort to such a person!) To be out of tune applies only to those who each time that they repeat a song, repeat it always differently and always offend the above-mentioned norms and never (except by some inexplicable accident) adhere to the original score, it being understood, of course, that they are not aware of it and on the contrary are firmly convinced each time that they are reproducing the score to the letter. In short, to be out of tune consists precisely in the inability to have any sort of relationship with the score, or to establish a steady point of reference in the great tossing sea of sounds.

These introductory remarks were necessary for the exposition of the following case which, I believe, is unique in the story of relationships between people.

There was a certain man in our town (I shall call him Giovanni) who lived only for music. Especially for lyric or operatic music or whatever you wish to call it. It was also claimed that he was endowed with a singularly melodious and robust voice, and he himself said that he devoted all of his time to singing. It was claimed, I repeat, since Giovanni, who was rich and completely independent, not only didn't make the slightest effort to have a public career, but did nothing to share his precious gift with anyone else. Indeed, not even his friends had ever heard him sing. But, in recompense, everyone had heard him in the opera houses, learnedly discussing the capacity of one singer or another as well as this or that note and how it had been delivered.

I had not always known Giovanni well, but gradually we became more and more friendly and so, one fine day, yielding to my reiterated insistence, he decided to grant me a concert in his house, which I had not yet been to. When I arrived there I quickly understood the reason for his constant reserve, as the reader will also immediately understand.

Giovanni had a wife, whom I now saw, a woman of great beauty and suavity. She was, so it seemed, very devoted to him. Blonde and very young, indeed almost a girl, and from a family which was without a doubt at least as noble as his own.

After having taken our places in the drawing room where the grand piano was located, Giovanni asked me to select a few from among the many arias—all from very well known operas—which he felt he knew best. So I selected some of those which were most familiar to me, in order to appreciate his art more fully, and then Giovanni, with his wife accompanying him, began to sing.

I was dumbfounded, unable at first to believe my own ears. In the exordium (it was the popular recitative from *Aida* which begins with the words *Se quel guerriero io fossi* . . .) I did not recognize a single note, nor could I grasp a single consecutive bar. Now, don't misunderstand me: it was not a matter here of the common off-key singing, ranging from a quarter to half a tone, to which amateur singers and people singing on the street have more or less accustomed us. It was a real jamboree of capriciously clashing sounds, which, believe me, not only had no relation whatsoever to the score but also did not bear the slightest relation to each other.

I had no reason to think that my memory of the tune had failed me, and besides, the accompaniment was there and must have meant something. I could do nothing but hope that my friend was following his personal counterpoint and that this would eventually establish a firm relationship between the notes of the score and those emitted by him. Alas, I soon realized that any given note of the score was always matched by a different note in the singing. And even calling them notes is a bit too much: they were something intermediate or adulterated which cannot be found in nature, that is, on the keyboard. Even if I had wished to judge those bellows as something entirely independent from the score—as an original composition or improvisation—I would have been immediately undeceived by a resumption of the melody (let us call it that), when from those lips, prettily pursed and almost smiling, issued sounds which were not only discordant and lacerating but brand-new. Giovanni's voice in itself was not at all harsh or insipid; and yet, used so badly, it could not help but be disagreeable.

After the first piece Giovanni absently asked me for my opinion and before I could reply he had started another, then a third, a fourth. . . . I had to be careful to hide my reactions since, as he sang, he kept looking at me.

At last, to bring their hospitable courtesy to a worthy close, he asked his wife to perform a few duets with him, including some ensemble singing, and she graciously consented. And here a new, unimaginable surprise awaited me, indeed the greatest surprise of the evening. It is hardly a matter for wonder, it is even completely natural that a person who sings out of tune, if he has no way of checking on himself, does not know that he is singing out of tune. But what follows is truly a matter for amazement.

I must explain that, during the entire exhibition, I had assiduously observed the young wife's face, trying to determine what she thought of all this, and all that I had seen was the rapt expression with which she continued to gaze at her husband. This, however, did not seem to imply a definite opinion. Now, as soon as she opened her mouth, I immediately realized that she sang as much out of tune as he did. And that wasn't all—and this is the astounding part—she sang out of tune precisely and identically in his fashion, as if to his direction, according to his inspiration, his mode, no matter how varying and momentary. I am at a loss to add anything else.

If I had needed proof, I would have been supplied with incontrovertible evidence by the "duo" of this first duet, and then by the others which followed. Well, he who wishes to refer back to the introductory remarks which I have set at the beginning of this story, will easily understand that two people who sing out of tune cannot, by definition, sing *together,* except by mere chance and for a single note, at least a single note at a time. Yet these two unfailingly agreed on each and every note, or whatever you might call them, and they sang entire pieces with such moving accord in their out-of-tuneness that I, amazed, consternated, dejected, let my shapely ears be lacerated almost willingly, meanwhile abandoning myself to philosophical reflections, half bitter, half comforting.

I can imagine the objection that will be raised. Could it not have been that although she was aware of how matters stood, out of an excess of devotion and so as not to hurt her husband, the woman was trying not to disillusion him and was making an effort to follow him in his distortions, proving by this that she possessed an especially subtle ear? But, leaving aside my presence, which would have frustrated such a plan, who, even among the most expert of singers, could have succeeded in reproducing the sounds emitted by Giovanni, which, as I have already said, bore no relation at all to the universally known notes and, during their emission, were continually different and varied? Besides, the seriousness and gravity with which she went about her singing was by no means ambiguous.

When all had ended, the moment came to express my opinion, and this time Giovanni stared straight in my face so that I would be forced to reveal my innermost feelings. Shifting nervously, as on a bed of thorns, how I managed I don't quite know, but I proffered the conventional compliments and got out of that house as fast as I could. He, however, was not deceived by my gracious words, for afterwards he barely answered my greeting, thus showing that he was not even touched by doubts as to the excellence of his art. Then we lost sight of each other altogether.

That evening, as I was going home, I brooded over the obscure designs of nature's cruelty which instills in one person a vivid passion for the things which he cannot do, while it fills another with dislike for those things he can do very well, and so on. Yet nature is also benevolent, because with one hand it gives back that which it has taken with the other (though one does not see why it took it away in the first place). After all, weren't those two perfectly happy? Of their incapacity, nearly segregated from the world as they were, they had not the slightest suspicion and thus could, with the purest bliss, far from any menace, give themselves up to their passion—so true is it that our real abilities do not at all make up the substance of our existence. And one would have to prove that theirs was indeed an incapacity in the absolute sense. So I came to the definite conclusion that they were not only *not* humiliated but rather openly favored by fate—with which I was therefore for the moment reconciled.

However, this thread of thoughts then lacked an end which fell into my hands recently although I had lost all memory of Giovanni and everything connected with him. I heard that his young wife had suddenly died: she had burst a vein in her breast while singing. So Giovanni has in his turn been plunged into the gloomiest grief. And while waiting to tie together these far-reaching reflections, nothing remains for us but to adopt the explanation of the poet: Giovanni can quite well say of himself and his Annabel Lee:

> "But we loved with a love that was more than love—
>   I and my Annabel Lee—
> With a love that the winged seraphs in Heaven
>   Coveted her and me.
>
> And this was the reason that, long ago,
>   In this kingdom by the sea,
> A wind blew out of a cloud, chilling
>   My beautiful Annabel Lee. . . ."

Let us hope that he may at least find the consolation—for some persons insufficient—that is mentioned further on in the poem.

The Private War of Private Jacob

Joe Haldeman

With each step your boot heel cracks through the sun-dried crust and your foot hesitates, drops through an inch of red talcum powder, and then you draw it back up with another crackle. Fifty men marching in a line through this desert and they sound like a big bowl of breakfast cereal.

Jacob held the laser projector in his left hand and rubbed his right in the dirt. Then he switched hands and rubbed his left in the dirt. The plastic handles got very slippery after you'd sweated on them all day long, and you didn't want the damn thing to squirt out of your grip when you were rolling and stumbling and crawling your way to the enemy, and you couldn't use the strap, noplace off the parade ground; goddamn slide-rule jockey figured out where to put it, too high, take the damn thing off if you could. Take the goddamn helmet off too, if you could. No matter you were safer with it on. They said. And they were pretty strict, especially about the helmets.

"Look happy, Jacob." Sergeant Melford was always all smile and bounce before a battle. During a battle, too. He smiled at the tanglewire and beamed at his men while they picked their way through it—if you go too fast you get tripped and if you go too slow you get burned—and he had a sad smile when one of his men got zeroed and a shriek a happy shriek when they first saw the enemy and glee when an enemy got zeroed and nothing but smiles smiles smiles through the whole sorry mess. "If he *didn't* smile, just once," young-old Addison told Jacob, a long time ago, "just once he cried or frowned, there would be fifty people waiting for the first chance to zero that son of a bitch." And Jacob asked why and he said, "You just take a good look inside yourself the next time you follow that crazy son of a bitch into hell and you come back and tell me how you felt about him."

Jacob wasn't stupid, that day or this one, and he did keep an inside eye on what was going on under his helmet. What old Sergeant Melford did for him was mainly to make him glad that he wasn't crazy too, and no matter how bad things got, at least Jacob wasn't enjoying it like that crazy laughing grinning old Sergeant Melford.

He wanted to tell Addison and ask him why sometimes you were really scared or sick and you would look up and see Melford laughing his crazy ass off, standing over some steaming roasted body, and you'd have to grin, too, was it just so insane horrible or? Addison might have been able to tell Jacob but Addison took a low one and got hurt bad in both legs and the groin and it was a long time before he came back and then he wasn't young-old any more but just old. And he didn't say much any more.

With both his hands good and dirty, for a good grip on the plastic handles, Jacob felt more secure and he smiled back at Sergeant Melford.

"Gonna be a good one, Sarge." It didn't do any good to say anything else, like it's been a long march and why don't we rest a while before we hit them, Sarge, or, I'm scared and sick and if I'm gonna die I want it at the very first, Sarge: no. Crazy old Melford would be down on his hunkers next to you and give you a couple of friendly punches and josh around and flash those white teeth until you were about to scream or run but instead you wound up saying, "Yeah, Sarge, gonna be a good one."

We most of us figured that what made him so crazy was just that he'd been in this crazy war so long, longer than anybody could remember anybody saying he remembered; and he never got hurt while platoon after platoon got zeroed out from under him by ones and twos and whole squads. He never got hurt and maybe that bothered him, not that any of us felt sorry for the crazy son of a bitch.

Wesley tried to explain it like this: "Sergeant Melford is an improbability locus." Then he tried to explain what a locus was and Jacob didn't really catch it, and he tried to explain what an improbability was, and that seemed pretty simple but Jacob couldn't see what it all had to do with math. Wesley was a good talker though, and he might have one day been able to clear it up but he tried to run through the tanglewire, you'd think not even a civilian would try to do that, and he fell down and the little metal bugs ate his face.

It was twenty or maybe twenty-five battles later, who keeps track, when Jacob realized that not only did old Sergeant Melford never get hurt, but he never killed any of the enemy either. He just ran around singing out orders and being happy and every now and then he'd shoot off his projector but he always shot high or low or the beam was too broad. Jacob wondered about it but by this time he was more afraid, in a way, of Sergeant Melford than he was of the enemy, so he kept his mouth shut and he waited for someone else to say something about it.

Finally Cromwell, who had come into the platoon only a couple of weeks after Jacob, noticed that Sergeant Melford never seemed to zero anybody and he had this theory that maybe the crazy old son of a bitch was a spy for the other side. They had fun talking about that for a while, and then Jacob told them about the old improbability locus theory, and one of the new guys said he sure is an imperturbable locust all right, and they all had a good laugh, which was good because Sergeant Melford came by and joined in after Jacob told him what was so funny, not about the improbability locus, but the old joke about how do you make a hormone? You don't pay her. Cromwell laughed like there was no tomorrow and for Cromwell there wasn't even any sunset, because he went across the perimeter to take a crap and got caught in a squeezer matrix.

The next battle was the first time the enemy used the drainer field, and of course the projectors didn't work and the last thing a lot of the men learned was that the light plastic stock made a damn poor weapon against a long knife, of which the enemy had plenty. Jacob lived because he got in a lucky kick, aimed for the groin but got the kneecap, and while the guy was hopping around trying to stay upright he dropped his knife and Jacob picked it up and gave the guy a new orifice, eight inches wide and just below the navel.

The platoon took a lot of zeros and had to fall back, which they did very fast because the tanglewire didn't work in a drainer field, either. They left Addison behind, sitting back against a crate with his hands in his lap and a big drooly red grin not on his face.

With Addison gone, no other private had as much combat time as Jacob. When they rallied back at the neutral zone, Sergeant Melford took Jacob aside and wasn't really smiling at all when he said: "Jacob, you know that now if anything happens to me, you've got to take over the platoon. Keep them spread out and keep them advancing, and most of all, keep them happy."

Jacob said, "Sarge, I can tell them to keep spread out and I think they will, and all of them know enough to keep pushing ahead, but how can I keep them happy when I'm never very happy myself, not when you're not around."

That smile broadened and turned itself into a laugh. You crazy old son of a bitch, Jacob thought and because he couldn't help himself, he laughed too. "Don't worry about that," Sergeant Melford said. "That's the kind of thing that takes care of itself when the time comes."

The platoon practiced more and more with knives and clubs and how to use your hands and feet but they still had to carry the projectors into combat because, of course, the enemy could turn off the drainer field whenever he wanted to. Jacob got a couple of scratches and a piece of his nose cut off, but the medic put some cream on it and it grew back. The enemy started using bows and arrows so the platoon had to carry shields, too, but that was't too bad after they designed one that fit right over the projector, held sideways. One squad learned how to use bows and arrows back at the enemy and things got as much back to normal as they had ever been.

Jacob never knew exactly how many battles he had fought as

a private, but it was exactly forty-one. And actually, he wasn't a private at the end of the forty-first.

Since they got the archer squad, Sergeant Melford had taken to standing back with them, laughing and shouting orders at the platoon and every now and then loosing an arrow that always landed on a bare piece of ground. But this particular battle (Jacob's forty-first) had been going pretty poorly, with the initial advance stopped and then pushed back almost to the archers; and then a new enemy force breaking out on the other side of the archers.

Jacob's squad maneuvered between the archers and the new enemy soldiers and Jacob was fighting right next to Sergeant Melford, fighting pretty seriously while old Melford just laughed his fool head off, crazy son of a bitch. Jacob felt that split-second funny feeling and ducked and a heavy club whistled just over his head and bashed the side of Sergeant Melford's helmet and sheared the top of his helmet off just as neat as you snip the end off a soft-boiled egg. Jacob fell to his knees and watched the helmet full of stuff twirl end over end in back of the archers and he wondered why there were little glass marbles and cubes inside the grey-blue blood-streaked mushy stuff and then everything just went.

Inside a mountain of crystal under a mountain of rock, a tiny piezoelectric switch, sixty-four molecules in a cube, flipped over to the OFF *position and the following transaction took place at just less than the speed of light:*

> UNIT 10011001011MELFORD ACCIDENTALLY DEACTIVATED.
> SWITCH UNIT 1101011100JACOB TO CATALYST STATUS.
> (SWITCHING COMPLETED)
> ACTIVATE AND INSTRUCT UNIT 1101011100JACOB.

and came back again just like that. Jacob stood up and looked around. The same old sun-baked plain, but everybody but him seemed to be dead. Then he checked and the ones that weren't obviously zeroed were still breathing a bit. And, thinking about it, he knew why. He chuckled.

He stepped over the collapsed archers and picked up Melford's bleedy skull-cap. He inserted the blade of a knife between the helmet and the hair, shorting out the induction tractor that held the helmet on the head and served to pick up and transmit signals. Letting the helmet drop to the ground, he carefully bore the grisly balding bowl over to the enemy's crapper. Knowing exactly where to look, he fished out all the bits and pieces of crystal and tossed them down the smelly hole. Then he took the unaugmented brain back to the helmet and put it back the way he had found it. He returned to his position by Melford's body.

The stricken men began to stir and a few of the most hardy wobbled to their hands and knees.

Jacob threw back his head and laughed and laughed.

The Library of Babel

Jorge Luis Borges

Translated by James E. Irby

By this art you may contemplate
the variation of the 23 letters . . .
The Anatomy of Melancholy,
part 2, sect. II, mem. IV

The universe (which others call the Library) is composed of an indefinite and perhaps infinite number of hexagonal galleries, with vast air shafts between, surrounded by very low railings. From any of the hexagons one can see, interminably, the upper and lower floors. The distribution of the galleries is invariable. Twenty shelves, five long shelves per side, cover all the sides except two; their height, which is the distance from floor to ceiling, scarcely exceeds that of a normal bookcase. One of the free sides leads to a narrow hallway which opens onto another gallery, identical to the first and to all the rest. To the left and right of the hallway there are two very small closets. In the first, one may sleep standing up; in the other, satisfy one's fecal necessities. Also through here passes a spiral stairway, which sinks abysmally and soars upwards to remote distances. In the hallway there is a mirror which faithfully duplicates all appearances. Men usually infer from this mirror that the Library is not infinite (if it really were, why this illusory duplication?); I prefer to dream that its polished surfaces represent and promise the infinite . . . Light is provided by some spherical fruit which bear the name of lamps. There are two, transversally placed, in each hexagon. The light they emit is insufficient, incessant.

Like all men of the Library, I have traveled in my youth; I have wandered in search of a book, perhaps the catalogue of catalogues; now that my eyes can hardly decipher what I write, I am preparing to die just a few leagues from the hexagon in which I was born. Once I am dead, there will be no lack of pious hands to throw me over the railing; my grave will be the fathomless air; my body will sink endlessly and decay and dissolve in the wind generated by the fall, which is infinite. I say that the Library is unending. The idealists argue that the hexagonal rooms are a necessary form of absolute space or, at least, of our intuition of space. They reason that a triangular or pentagonal room is inconceivable. (The mystics claim that their ecstasy reveals to them a circular chamber containing a great circular book, whose spine is continuous and which follows the complete circle of the walls; but their testimony is suspect; their words, obscure. This cyclical book is God.) Let it suffice now for me to repeat the classic dictum: *The Library is a sphere whose exact center is any one of its hexagons and whose circumference is inaccessible.*

There are five shelves for each of the hexagon's walls; each shelf contains thirty-five books of uniform format; each book is of four hundred and ten pages; each page, of forty lines, each line, of some eighty letters which are black in color. There are also letters on the spine of each book; these letters do not indicate or prefigure what the pages will say. I know that this incoherence at one time seemed mysterious. Before summarizing the solution (whose discovery, in spite of its tragic projections, is perhaps the capital fact in history) I wish to recall a few axioms.

First: The Library exists *ab aeterno.* This truth, whose immediate corollary is the future eternity of the world, cannot be placed in doubt by any reasonable mind. Man, the imperfect librarian, may be the product of chance or of malevolent demiurgi; the universe, with its elegant endowment of shelves, of enigmatical volumes, of inexhaustible stairways for the traveler and latrines for the seated librarian, can only be the work of a god. To perceive the distance between the divine and the human, it is enough to compare these crude wavering symbols which my fallible hand scrawls on the cover of a book, with the organic letters inside: punctual, delicate, perfectly black, inimitably symmetrical.

Second: *The orthographical symbols are twenty-five in number.** This finding made it possible, three hundred years ago, to formulate a general theory of the Library and solve satisfactorily the problem which no conjecture had deciphered: the formless and chaotic nature of almost all the books. One which my father saw in a hexagon on circuit fifteen ninety-four was made up of the letters MCV, perversely repeated from the first line to the last. Another (very much consulted in this area) is a mere labyrinth of letters, but the next-to-last page says *Oh time thy pyramids.* This much is already known: for every sensible line of straightforward statement, there are leagues of senseless cacophonies, verbal jumbles and incoherences. (I know of an uncouth region whose librarians repudiate the vain and superstitious custom of finding a meaning in books and equate it with that of finding a meaning in dreams or in the chaotic lines of one's palm . . . They admit that the inventors of this writing imitated the twenty-five natural symbols, but maintain that this application is accidental and that the books signify nothing in themselves. This dictum, we shall see, is not entirely fallacious.)

For a long time it was believed that these impenetrable books corresponded to past or remote languages. It is true that the most ancient men, the first librarians, used a language quite different from the one we now speak; it is true that a few miles to the right the tongue is dialectal and that ninety floors farther up, it is incomprehensible. All this, I repeat, is true, but four hundred and ten pages of inalterable MCV's cannot correspond to any lan-

* The original manuscript does not contain digits or capital letters. The punctuation has been limited to the comma and the period. These two signs, the space and the twenty-two letters of the alphabet are the twenty-five symbols considered sufficient by this unknown author. (*Editor's note.*)

guage, no matter how dialectal or rudimentary it may be. Some insinuated that each letter could influence the following one and that the value of MCV in the third line of page 71 was not the one the same series may have in another position on another page, but this vague thesis did not prevail. Others thought of cryptographs; generally, this conjecture has been accepted, though not in the sense in which it was formulated by its originators.

Five hundred years ago, the chief of an upper hexagon* came upon a book as confusing as the others, but which had nearly two pages of homogeneous lines. He showed his find to a wandering decoder who told him the lines were written in Portuguese; others said they were Yiddish. Within a century, the language was established: a Samoyedic Lithuanian dialect of Guarani, with classical Arabian inflections. The content was also deciphered: some notions of combinative analysis, illustrated with examples of variation with unlimited repetition. These examples made it possible for a librarian of genius to discover the fundamental law of the Library. This thinker observed that all the books, no matter how diverse they might be, are made up of the same elements: the space, the period, the comma, the twenty-two letters of the alphabet. He also alleged a fact which travelers have confirmed: *In the vast Library there are no two identical books.* From these two incontrovertible premises he deduced that the Library is total and that its shelves register all the possible combinations of the twenty-odd orthographical symbols (a number which, though extremely vast, is not infinite): in other words, all that it is given to express, in all languages. Everything: the minutely detailed history of the future, the archangels' autobiographies, the faithful catalogue of the Library, thousands and thousands of false catalogues, the demonstration of the fallacy of those catalogues, the demonstra-

* Before, there was a man for every three hexagons. Suicide and pulmonary diseases have destroyed that proportion. A memory of unspeakable melancholy: at times I have traveled for many nights through corridors and along polished stairways without finding a single librarian.

tion of the fallacy of the true catalogue, the Gnostic gospel of Basilides, the commentary on that gospel, the commentary on the commentary on that gospel, the true story of your death, the translation of every book in all languages, the interpolations of every book in all books.

When it was proclaimed that the Library contained all books, the first impression was one of extravagant happiness. All men felt themselves to be the masters of an intact and secret treasure. There was no personal or world problem whose eloquent solution did not exist in some hexagon. The universe was justified, the universe suddenly usurped the unlimited dimensions of hope. At that time a great deal was said about the Vindications: books of apology and prophecy which vindicated for all time the acts of every man in the universe and retained prodigious arcana for his future. Thousands of the greedy abandoned their sweet native hexagons and rushed up the stairways, urged on by the vain intention of finding their Vindication. These pilgrims disputed in the narrow corridors, proffered dark curses, strangled each other on the divine stairways, flung the deceptive books into the air shafts, met their death cast down in a similar fashion by the inhabitants of remote regions. Others went mad . . . The Vindications exist (I have seen two which refer to persons of the future, to persons who perhaps are not imaginary) but the searchers did not remember that the possibility of a man's finding his Vindication, or some treacherous variation thereof, can be computed as zero.

At that time it was also hoped that a clarification of humanity's basic mysteries—the origin of the Library and of time—might be found. It is verisimilar that these grave mysteries could be explained in words: if the language of philosophers is not sufficient, the multiform Library will have produced the unprecedented language required, with its vocabularies and grammars. For four centuries now men have exhausted the hexagons . . . There are official searchers, *inquisitors*. I have seen them in the performance of their function: they always arrive extremely tired from their journeys; they speak of a broken stairway which almost killed them; they talk with the librarian of galleries and stairs; sometimes they pick up the nearest volume and leaf through it, looking for infamous words. Obviously, no one expects to discover anything.

As was natural, this inordinate hope was followed by an excessive depression. The certitude that some shelf in some hexagon held precious books and that these precious books were inaccessible, seemed almost intolerable. A blasphemous sect suggested that the searches should cease and that all men should juggle letters and symbols until they constructed, by an improbable gift of chance, these canonical books. The authorities were obliged to issue severe orders. The sect disappeared, but in my childhood I have seen old men who, for long periods of time, would hide in the latrines with some metal disks in a forbidden dice cup and feebly mimic the divine disorder.

Others, inversely, believed that it was fundamental to eliminate useless works. They invaded the hexagons, showed credentials which were not always false, leafed through a volume with displeasure and condemned whole shelves: their hygienic, ascetic furor caused the senseless perdition of millions of books. Their name is execrated, but those who deplore the "treasures" destroyed by this frenzy neglect two notable facts. One: the Library is so enormous that any reduction of human origin is infinitesimal. The other: every copy is unique, irreplaceable, but (since the Library is total) there are always several hundred thousand imperfect facsimiles: works which differ only in a letter or a comma. Counter to general opinion, I venture to suppose that the consequences of the Purifiers' depredations have been exaggerated by the horror these fanatics produced. They were urged on by the delirium of trying to reach the books in the Crimson Hexagon: books whose format is smaller than usual, all-powerful, illustrated and magical.

We also know of another superstition of that time: that of the Man of the Book. On some shelf in some hexagon (men reasoned) there must exist a book which is the formula and perfect compendium *of all the rest:* some librarian has gone though it and he is analogous to a god. In the language of this zone vestiges of this remote functionary's cult still persist. Many wandered in search of Him. For a century they exhausted in vain the most varied areas. How could one locate the venerated and secret hexagon which housed Him? Someone proposed a regressive method: To locate book A, consult first a book B which indicates A's position; to locate book B, consult first a book C, and so on to infinity . . . In adventures such as these, I have squandered and wasted my years. It does not seem unlikely to me that there is a total book on some shelf of the universe;* I pray to the unknown gods that a man—just one, even though it were thousands of years ago!—may have examined and read it. If honor and wisdom and happiness are not for me, let them be for others. Let heaven exist, though my place be in hell. Let me be outraged and annihilated, but for one instant, in one being, let Your enormous Library be justified. The impious maintain that nonsense is normal in the Library and that the reasonable (and even humble and pure coherence) is an almost miraculous exception. They speak (I know) of the "feverish Library whose chance volumes are constantly in danger of changing into others and affirm, negate and confuse everything like a delirious divinity." These words, which not only denounce the disorder but exemplify it as well, notoriously prove their authors' abominable taste and desperate ignorance. In truth, the Library includes all verbal structures, all variations permitted by the twenty-five orthographical symbols, but not a single example of absolute nonsense. It is useless to observe that the best volume of the many hexagons under my administration is entitled *The Combed Thunderclap* and another *The Plaster Cramp* and another *Axaxaxas mlö*. These phrases, at first glance incoherent, can no doubt be justified in a cryptographical or allegorical manner; such a justification is verbal and, *ex hypothesi*, already figures in the Library. I cannot combine some characters

* I repeat: it suffices that a book be possible for it to exist. Only the impossible is excluded. For example: no book can be a ladder, although no doubt there are books which discuss and negate and demonstrate this possibility and others whose structure corresponds to that of a ladder.

dhcmrlchtdj

which the divine Library has not foreseen and which in one of its secret tongues do not contain a terrible meaning. No one can articulate a syllable which is not filled with tenderness and fear, which is not, in one of these languages, the powerful name of a god. To speak is to fall into tautology. This wordy and useless epistle already exists in one of thc thirty volumes of the five shelves of one of the innumerable hexagons—and its refutation as well. (An *n* number of possible languages use the same vocabulary; in some of them, the symbol *library* allows the correct definition *a ubiquitous and lasting system of hexagonal galleries,* but *library* is *bread* or *pyramid* or anything else, and these seven words which define it have another value. You who read me, are You sure of understanding my language?)

The methodical task of writing distracts me from the present state of men. The certitude that everything has been written negates us or turns us into phantoms. I know of districts in which the young men prostrate themselves before books and kiss their pages in a barbarous manner, but they do not know how to decipher a single letter. Epidemics, heretical conflicts, peregrinations which inevitably degenerate into banditry, have decimated the population. I believe I have mentioned the suicides, more and more frequent with the years. Perhaps my old age and fearfulness deceive me, but I suspect that the human species—the unique species—is about to be extinguished, but the Library will endure: illuminated, solitary, infinite, perfectly motionless, equipped with precious volumes, useless, incorruptible, secret.

I have just written the word "infinite." I have not interpolated this adjective out of rhetorical habit; I say that it is not illogical to think that the world is infinite. Those who judge it to be limited postulate that in remote places the corridors and stairways and hexagons can conceivably come to an end—which is absurd. Those who imagine it to be without limit forget that the possible number of books does have such a limit. I venture to suggest this solution to the ancient problem: *The Library is un-*

limited and cyclical. If an eternal traveler were to cross it in any direction, after centuries he would see that the same volumes were repeated in the same disorder (which, thus repeated, would be an order: the Order). My solitude is gladdened by this elegant hope.*

* Letizia Alvarez de Toledo has observed that this vast Library is useless: rigorously speaking, *a single volume* would be sufficient, a volume of ordinary format, printed in nine or ten point type, containing an infinite number of infinitely thin leaves. (In the early seventeenth century, Cavalieri said that all solid bodies are the superimposition of an infinite number of planes.) The handling of this silky vade mecum would not be convenient: each apparent page would unfold into other analogous ones; the inconceivable middle page would have no reverse.

Enantiomorphosis (A Natural History of Mirrors)

Christian Bök

1.

The Catoptriarchs (c. 711–777 A.D.), a Slavonic sect of Christian gnostics, advocated the Enantiomorphic Heresy, which declared that all earthly existence was but a fleeting reflection in a looking-glass unveiled in the gardens of the Heavenly Father. Catoptriarchs expressed their *contemptus mundi* by refusing to gaze upon the world unless it was reflected in a mirror, and any disciple of this sect was easily identified by his periscopic obsession with a sheet of silver plating, which he always carried with him in one hand, and from which his glance never strayed, even when riding on horseback. Catoptriarchs earned erroneous repute for their narcissism and suffered persecution by the official clergy in Constantinople. Manuscripts written by the Catoptriarchs were almost all destroyed because the Slavonic mirror-writing in such texts was entirely illegible and thus often mistaken for Satanic speech by monastic archivists, who burnt such books for the blasphemy of their form, not of their content.

René-Just Haüy (1743–1822), the French crystallographer, numbered among the inventory of his scientific library an incunabulum thought to be the Latin transcription of a manuscript originally written by a Catoptriarch. The Latin was enciphered by an anonymous monk, who wrote using an alphabet inspired by the diffraction patterns that sunbeams make when passing through either prisms of rock quartz or layered panes of glass (see FIGURE 3.8).

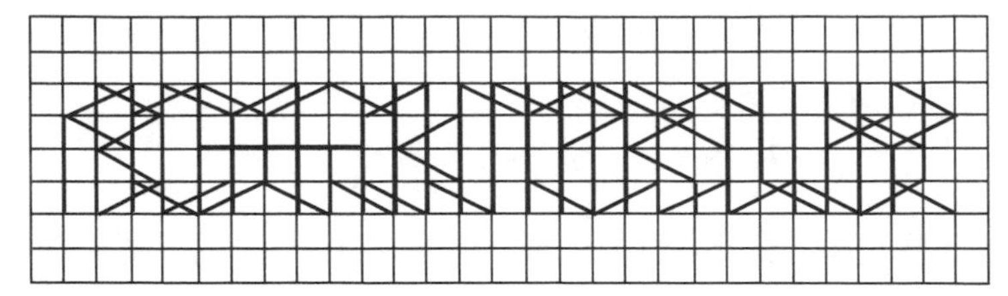

Figure 3.8
An alphabet for plotting the transmission of light-rays through a transparent medium.

The French Revolution saw the destruction of this book in the fire at the library of Haüy, when arsonists stormed the streets and indiscriminately burned any property that might even suggest the lifestyle of an aristocrat.

Documentary evidence about the Catoptriarchs now exists only in the fragmentary marginalia of private diaries written by Haüy after the fire, diaries in which he not only recounts from memory the above anecdotes, but also reconstructs the prismatic alphabet of his lost text.

2.

De Speculum Oraculum (c. 1230 A.D.), the final volume in a trilogy about medieval glassblowing, now resides in the repository of forbidden books at the Vatican, since the text allegedly describes heretical techniques for predicting the advent of the apocalypse. Written on scorched, vellum pages, interleaved between polished, steel plates, all bound together by solid, glass clamps, one at each corner, the text discourses upon the arts of both catoptromancy and crystallomancy, two genres of soothsaying that allow a clairvoyant to foresee the future, to witness the end of the world, by gazing into the hypnotic abysses of either mirrors or crystals.

De Speculum purports to be the *vade mecum*, the handbook, for a secret guild of German glassblowers supposedly able to manufacture a mirror that can, when smashed, depict a multiplicity of destinies in its mosaic of fragments, each piece containing a different image, a single scene, from the life of the person last reflected in the glass. The diverse images among the random shards do not appear to have any coherent order so that the clairvoyant must interpret their proper relations within the arrangement. The clairvoyant constructs a narrative from the mosaic, whose images stay visible so long as the pattern of their shattering goes undisturbed, for even a slight movement of the mosaic causes them to disappear, never to return.

De Speculum concludes with a natural history of crystals, complete with diagrams, the text alluding to a theory of optics, in which the pathways of light through a jewel embody the pathways of thought through the mind: optical reflection provides an allegory for cerebral reflection.

De Speculum goes on to warn that gazing into a crystal for too long can risk causing the soul of the clairvoyant to become trapped irrevocably within the faceted depths of the gem. Imprisonment of the soul in this way causes the clairvoyant to go insane, and destruction of the gem results in the immediate death of the imprisoned victim. The text in fact predicts that the world ends when men lose their minds by emptying their memories into mirrors in the same way that scribes write messages upon a page.

3.

Christian Weiss (1780–1856), the German crystallographer, recounts in a loveletter to his mistress the following, curious anecdote about a discovery made by him during his research on crystalline symmetry at the Berlin Academy.

Weiss reports purchasing a medieval treatise on the use of mirrors in the game of chess, a treatise found by him at an antiquarian bookstall on the verge of the ghetto. Weiss writes that the prolegomenon of the book chronicles the apocryphal experiments of a Saracen alchemist reputed to have made items vanish by placing them between a pair of looking-glasses in the court of his patron, the Caliph.

Double mirrors facing each other can, according to this book, trap the spirits of the dead who pass between them; moreover, any living person who has no soul can actually step into either one of the mirrors as if it were an open door and thus walk down the illusory corridor that appears to recede forever into the depths of the glass by virtue of one mirror reflecting itself in the other. The walls of such a corridor are said to be made from invulnerable panes of crystal, beyond which lies a nullified dimension of such complexity that to view it is surely to go insane. The book also explains at length that, after an eternity of walking down such a corridor, a person eventually exits from the looking-glass opposite to the one first entered.

Weiss speculates that a soulless man might carry another pair of mirrors into such a corridor, thereby producing a hallway at right angles to the first one, and of course this procedure might be performed again and again in any of the corridors until an endless labyrinth of glass has been erected inside the first pair of mirrors, each mirror opening onto an extensive grid of crisscrossing hallways, some of which never intersect, despite their lengths being both infinite and perpendicular. Weiss expresses his own misgivings about becoming hopelessly lost while exploring such a maze, and he wonders what happens to the prisoner if the initial pair of mirrors are disturbed so that they no longer reflect each other, thus suddenly obliterating the fragile foundation upon which the entire maze rests.

Weiss never divulges the title of this treatise, and since no one has ever corroborated the existence of such a book, scholars have simply dismissed this bibliophilic anecdote as a fiction designed to entertain his mistress, a countess rumoured to grow easily bored with lovers unable to appease her intellectual appetite for enigmas, puzzles, and rebuses.

4.

Rev. Charles Lutwidge Dodgson (1832–1898), the British mathematician, documented his initial experimentation with hallucinogenic drugs in an unpublished monograph torn from his private journal and tipped into a single copy of the first edition of his treatise on geometry, *Euclid and His Modern Rivals,* a copy now thought to be in the possession of a private collector in Luxembourg.

Dodgson reported that, while reading the book *Al Aaraaf* during his studies at Christ Church in Oxford, he grew more curious about altered states of mental awareness and thus decided to visit a pharmacy where he acquired a dose of crystallized opiates in the hope that he too might experience within the privacy of his own garret the visions of the poetry. Dodgson wrote that, shortly after administering the drug in an infusion of laudanum, he began feeling an ominous anxiety while contemplating the halo of light-rays emanating from a gaslamp in his room, for the rays seemed to grow steadily more distinct, more refined, attenuating themselves into long, slender needles honed to a divine acuity, the lamp transmuting into what he described as "a thistle of illumination, radiating acicular beams of searing energy that pierced through any object in their path." Distressed, he stood up suddenly from his desk and accidentally knocked over a looking-glass, only to see it topple in slow motion onto the floor, where the mirror burst into smithereens.

Dodgson claimed that, to his amazement, the glittering fragments of the broken mirror floated up one by one into the air, each a spinning prism that began to orbit elliptically around his head and body at an accelerated rate until all the shards formed a kind of kaleidoscopic shield that deflected the sharpened beams of gaslight. Dodgson, euphoric, described himself later as "a tree surrounded by its nimbus of silver leaves when the wind shivers them to life." Mesmerized, he attempted to step out through this cyclonic barrier of flying shrapnel, only to faint when the light-rays pierced him at every point and the bladed glass-shards sliced him to shreds. Dodgson awoke, uninjured, hours later beside his broken mirror and promptly recorded the hallucination in his journal, but then decided to keep his experience secret for fear of being diagnosed as deranged by his academic colleagues, who thought him a bit queer for befriending pubescent girls by composing for them nonsensical verse.

5.

Jacques Lacan (1901–1981), the French psychoanalyst, first delivered his seminal essay "The Mirror Stage as Formative of the

Function of the I" on July 17, 1946 while attending the 16th International Congress of Psychoanalysis in Zürich. Lacan upheld the idea that an infant in the first months of its life could only apprehend the experience of both self and other via *méconnaissance*, the child misconstruing itself as other when seeing its own reflection in a mirror for the first time. Lacan went on to argue that this developmental stage induced a crisis of alienation, since the subject defined itself through an external image, whose ideality could never coincide exactly with the identity of the child, and because of this irreconcilable distance between the subject and its reflection, the child not only admired its own image for providing a coherent representation of the self, but also despised its own image for withholding a representation congruent with the self. Lacan concluded that this psychodrama marked the transition of a subject from a state of lack to a state of desire.

Le Monde reported in its morning edition on the same day as this academic symposium that Dr. Jules Verrier, a clinical psychiatrist, had gone insane while trying to escpe from an elevator that had trapped him alone overnight between the floors of his office-building. Verrier had apparently screamed for almost seven hours straight, all the while clawing away at the mirrored interior of his compartment, before firemen finally rescued him and then escorted him away under physical restraint. Sources close to the scene testified later that the psychiatrist had behaved quite hysterically, insisting to authorities that he had become involved in a case of mistaken identity, that he was the wrong person to be taken into custody, since his reflection in the mirrored wall of the elevator had somehow traded places with him. His reflection had suddenly refused to imitate him and had stepped out of its frame so as to force him to occupy the space vacated in the glass.

Verrier was committed a day later to l'Hôpital Général in Paris, where doctors tried to console him by telling him that, obviously, he was not imprisoned in a mirror, but he responded to treatment by yelling in frustrated rage that nobody understood him, for he insisted that this world was itself the very world beyond the mirror, a world from which his reflection had escaped, tearing itself free so as to wreak havoc upon the true reality. Medical authorities at the asylum diagnosed the patient as a paranoid catoptrophobiac, but confessed ironically that they now preferred to exit buildings via the stairs.

Notes

ENANTIOMORPHIC projections occur in crystals when one
crystal mirrors the structure of another so that both
create a pair of optical antipodes, each one defined
as the *enantiomer* of the other, depending upon the axis
of reflection, be it either vertical or horizontal:
for example, a vertical axis makes enantiomers of not
only *b* and *d*, but also *p* and *q*, just as a horizontal
axis makes enantiomers of not only *p* and *b*, but also
q and *d*; however, no axis can ever make enantiomers
of *b* and *q*, just as no axis can ever make enantiomers
of *d* and *p*. Words can form enantiomorphic projections
of each other only when one crystal translates itself
into the other through reflection. The enantiomorphic
projection often defines the relationship between two
components of a crystal whose infrastructure exhibits
interpenetrant twinning: for example, *w* takes shape
at the very moment when *v* twins with its enantiomer
through a vertical axis, just as X takes shape at the
very moment when *v* twins with its enantiomer through
a horizontal axis. Such catoptric symmetries underlie
the taxonomic categorization of all crystalline forms.

Mirrors have historically played a significant role
in advancing the science of crystallography, insofar
as the optical properties of mirrors provided
the mathematical means for identifying a crystalline
structure on the basis of its internal symmetries.
The class of symmetry to which a crystal belongs is
determined by slicing a crystal conceptually along
diverse axes with a mirrored blade. The reflections
in the blade define the degree to which the crystal
is symmetrical with itself. The term "enantiomorphic"
applies also to more complex, crystalline structures,
such as palindromes: for example, the phrase *mirror*
rim reveals a sequential symmetry, in which the order
of letters before the letter *o* reiterates in reverse
the order of letters after the letter *o*, with each
letter expressing an enantiomorphic quality in its
own structure, the doubled *r*, doubled, the letters *m*,
i, and *o*. each symmetrical through a central, vertical
axis, the gap between the two words, a flaw in the gem.

MIRRORS INDUCE DYSLEXIA

Schwarzschild Radius

Connie Willis

"When a star collapses, it sort of falls in on itself." Travers curved his hand into a semicircle and then brought the fingers in, "and sometimes it reaches a kind of point of no return where the gravity pulling in on it is stronger than the nuclear and electric forces, and when it reaches that point nothing can stop it from collapsing and it becomes a black hole." He closed his hand into a fist. "And that critical diameter, that point where there's no turning back, is called the Schwarzschild radius." Travers paused, waiting for me to say something.

He had come to see me every day for a week, sitting stiffly on one of my chairs in an unaccustomed shirt and tie, and talked to me about black holes and relativity, even though I taught biology at the university before my retirement, not physics. Someone had told him I knew Schwarzschild, of course.

"The Schwarzschild radius?" I said in my quavery, old man's voice, as if I could not remember ever hearing the phrase before, and Travers looked disgusted. He wanted me to say, "The Schwarzschild radius! Ah, yes, I served with Karl Schwarzschild on the Russian front in World War I!" and tell him all about how he had formulated his theory of black holes while serving with the artillery, but I had not decided yet what to tell him. "The event horizon," I said.

"Yeah. It was named after Schwarzschild because he was the one who worked out the theory," Travers said. He reminded me of Muller with his talk of theories. He was the same age as Muller, with the same shock of stiff yellow hair and the same insatiable curiosity, and perhaps that was why I let him come every day to talk to me, though it was dangerous to let him get so close.

"I have drawn up a theory of the stars," Muller says while we warm our hands over the Primus stove so that they will get enough feeling in them to be able to hold the liquid barretter without dropping it. "They are not balls of fire, as the scientists say. They are frozen."

"How can we see them if they are frozen?" I say. Muller is insulted if I do not argue with him. The arguing is part of the theory.

"Look at the wireless!" he says, pointing to it sitting disemboweled on the table. We have the back off the wireless again, and in the barretter's glass tube is a red reflection of the stove's flame. "The light is a reflection off the ice of the star."

"A reflection of what?"

"Of the shells, of course."

I do not say that there were stars before there was this war, because Muller will not have an answer to this, and I have no desire to destroy his theory, and besides, I do not really believe there was a time when this war did not exist. The star shells have always exploded over the snow-covered craters of No Man's Land, shattering in a spray of white and red, and perhaps Muller's theory is true.

"At that point," Travers said, "at the event horizon, no more information can be transmitted out of the black hole because gravity has become so strong, and so the collapse appears frozen at the Schwarzschild radius."

"Frozen," I said, thinking of Muller.

"Yeah. As a matter of fact, the Russians call black holes 'frozen stars.' You were at the Russian front, weren't you?"

"What?"

"In World War I."

"But the star doesn't really freeze," I said. "It goes on collapsing."

"Yeah, sure," Travers said. "It keeps collapsing in on itself until even the atoms are stripped of their electrons and there's nothing left except what they call a naked singularity, but we can't see past the Schwarzschild radius, and nobody inside a black hole can tell us what it's like in there because they can't get messages out, so nobody can ever know what it's like inside a black hole."

"I know," I said, but he didn't hear me.

He leaned forward. "What was it like at the front?"

It is so cold we can only work on the wireless a few minutes at a time before our hands stiffen and grow clumsy, and we are afraid of dropping the liquid barretter. Muller holds his gloves over the Primus stove and then puts them on. I jam my hands in my ice-stiff pockets.

We are fixing the wireless set. Eisner, who had been delivering messages between the sectors, got sent up to the front when he could not fix his motorcycle. If we cannot fix the wireless we will cease to be telegraphists and become soldiers and we will be sent to the front lines.

We are already nearly there. If it were not snowing we could see the barbed wire and pitted snow of No Man's Land, and the big Russian coalboxes sometimes land in the communication trenches. A shell hit our wireless hut two weeks ago. We are ahead of our own artillery lines, and some of the shells from our guns fall on us, too, because the muzzles are worn out. But it is not the front, and we guard the liquid barretter with our lives.

"Eisner's unit was sent up on wiring fatigue last night," Muller says, "and they have not come back. I have a theory about what happened to them."

"Has the mail come?" I say, rubbing my sore eyes and then putting my cold hands immediately back in my pockets. I must get some new gloves, but the quartermaster has none to issue. I have written my mother three times to knit me a pair, but she has not sent them yet.

"I have a theory about Eisner's unit," he says doggedly. "The Russians have a magnet that has pulled them into the front."

"Magnets pull iron, not people," I say.

I have a theory about Muller's theories. Littering the communications trenches are things that the soldiers going up to the front have discarded: water bottles and haversacks and bayonets. Hans and I sometimes tried to puzzle out why they would discard such important things.

"Perhaps they were too heavy," I would say, though that did not explain the bayonets or the boots.

"Perhaps they know they are going to die," Hans would say, picking up a helmet.

I would try to cheer him up. "My gloves fell out of my pocket yesterday when I went to the quartermaster's. I never found them. They are in this trench somewhere."

"Yes," he would say, turning the helmet round and round in his hands, "perhaps as they near the front, these things simply drop away from them."

My theory is that what happens to the water bottles and helmets and bayonets is what has happened to Muller. He was a student in university before the war, but his knowledge of science and his intelligence have fallen away from him, and now we are so close to the front, all he has left are his theories. And his curiosity, which is a dangerous thing to have kept.

"Exactly. Magnets pull iron, but *they* were carrying barbed wire!" he says triumphantly, "and so they were pulled in to the magnet."

I put my hands practically into the Primus flame and rub them together, trying to get rid of the numbness. "We had better get the barretter in the wireless again or this magnet of yours will suck it in, too."

I go back to the wireless. Muller stays by the stove, thinking about his magnet. The door bangs open. It is not a real door, only an iron humpie tied to the beam that reinforces the dugout and held with a wedge, and when someone pushes against it, it flies inward, bringing the snow with it.

Snow swirls in, and light, and the sound from the front, a low rumble like a dog growling. I clutch the liquid barretter to my chest and Muller flings himself over the wireless as if it were a wounded comrade. Someone bundled in a wool coat and mittens, with a wool cap pulled over his ears, stands silhouetted against the reddish light in the doorway, blinking at us.

"Is Private Rottschieben here? I have come to see him about his eyes," he says, and I see it is Dr. Funkenheld.

"Come in and shut the door," I say, still carefully protecting the liquid barretter, but Muller has already jammed the metal back against the beam.

"Do you have news?" Muller says to the doctor, eager for new facts to spin his theories from. "Has the wiring fatigue come back? Is there going to be a bombardment tonight?"

Dr. Funkenheld takes off his mittens. "I have come to examine your eyes," he says to me. His voice frightens me. All through the war he has kept his quiet bedside voice, speaking to the wounded in the dressing station and at the stretcher bearer's posts as if they were in his surgery in Stuttgart, but now he sounds agitated and I am afraid it means a bombardment is coming and he will need me at the front.

When I went to the dressing station for medicine for my eyes, I foolishly told him I had studied medicine with Dr. Zuschauer in Jena. Now I am afraid he will ask me to assist him, which will mean going up to the front. "Do your eyes still hurt?" he says.

I hand the barretter to Muller and go over to stand by the lantern that hangs from a nail in the beam.

"I think he should be invalided home, Herr Doktor," Muller says. He knows it is impossible, of course. He was at the wireless the day the message came through that no one was to be invalided out for frostbite or "other noncontagious diseases."

"Can you find me a better light?" the doctor says to him.

Muller's curiosity is so strong that he cannot bear to leave any place where something interesting is happening. If he went up to the front I do not think he would be able to pull himself away, and now I expect him to make some excuse to stay, but I have forgotten that he is even more curious about the wiring fatigue. "I will go see what has happened to Eisner's unit," he says, and opens the door. Snow flies in, as if it had been beating against the door to get in, and the doctor and I have to push against the door to get it shut again.

"My eyes have been hurting," I say, while we are still pushing the metal into place, so that he cannot ask me to assist him. "They feel like sand has gotten into them."

"I have a patient with a disease I do not recognize," he says. I am relieved, though disease can kill us as easily as a trench mortar. Soldiers die of pneumonia and dysentery and blood poisoning every day in the dressing station, but we do not fear it the way we fear the front.

"The patient has fever, excoriated lesions, and suppurating bullae," Dr. Funkenheld says.

"Could it be boils?" I say, though of course he would recognize something so simple as boils, but he is not listening to me, and I realize that it is not a diagnosis from me that he has come for.

"The man is a scientist, a Jew named Schwarzschild, attached to the artillery," he says, and because the artillery are even farther back from the front lines than we are, I volunteer to go and look at the patient, but he does not want that either.

"I must talk to the medical headquarters in Bialystok," he says.

"Our wireless is broken," I say, because I do not want to have to tell him why it is impossible for me to send a message for him. We are allowed to send only military messages, and they must be sent in code, tapped out on the telegraph key. It would take hours to send his message, even if it were possible. I hold up the dangling wire. "At any rate, you must clear it with the commandant," but he is already writing out the name and address on a piece of paper, as if this were a telegraph office.

"You can send the message when you get the wireless fixed. I have written out the symptoms."

I put the back on the wireless. Muller comes in, kicking the door open, and snow flies everywhere, picking up Dr. Funkenheld's message and sending it circling around the dugout. I catch it before it spirals into the flame of the Primus stove.

"The wiring fatigue was pinned down all night," Muller says, setting down a hand lamp. He must have gotten it from the dressing station. "Five of them frozen to death, the other eight have frostbite. The commandant thinks there may be a bombardment tonight." He does not mention Eisner, and he does not say what has happened to the rest of the thirty men in Eisner's unit, though I know. The front has gotten them. I wait, holding the message in my stiff fingers, hoping Dr. Funkenheld will say, "I must go attend to their frostbite."

"Let me examine your eyes," the doctor says, and shows Muller how to hold the hand lamp. Both of them peer into my eyes. "I have an ointment for you to use twice daily," he says, getting a flat jar out of his bag. "It will burn a little."

"I will rub it on my hands then. It will warm them," I say, thinking of Eisner frozen at the front, still holding the roll of barbed wire, perhaps.

He pulls my bottom eyelid down and rubs the ointment on with his little finger. It does not sting, but when I have blinked it into my eye, everything has a reddish tinge. "Will you have the wireless fixed by tomorrow?" he says.

"I don't know. Perhaps."

Muller has not put down the hand lamp. I can see by its light that he has forgotten all about the wiring fatigue and the Russian magnet and is wondering what the doctor wants with the wireless.

The doctor puts on his mittens and picks up his bag. I realize too late I should have told him I would send the message in exchange for them. "I will come check your eyes tomorrow," he says and opens the door to the snow. The sound of the front is very close.

As soon as he is gone, I tell Muller about Schwarzschild and the message the doctor wants to send. He will not let me rest until I have told him, and we do not have time for his curiosity. We must fix the wireless.

"If you were on the wireless, you must have sent messages for Schwarzschild," Travers said eagerly. "Did you ever send a message to Einstein? They've got the letter Einstein sent to him after he wrote him his theory, but if Schwarzschild sent him some kind of message, too, that would be great. It would make my paper."

"You said that no message can escape a black hole?" I said. "But they could escape a collapsing star. Is that not so?"

"Okay," Travers said impatiently and made his fingers into a semicircle again. "Suppose you have a fixed observer over here." He pulled his curved hand back and held the forefinger of his other hand up to represent the fixed observer, "and you have somebody in the star. Say when the star starts to collapse, the person in it shines a light at the fixed observer. If the star hasn't reached the Schwarzschild radius, the fixed observer will be able to see the light, but it will take longer to reach him because the gravity of the black hole is pulling on the light, so it will seem as if time on the star has slowed down and the wavelengths will have been lengthened, so the light will be redder. Of course that's just a thought problem. There couldn't really be anybody in a collapsing star to send the messages."

"We sent messages," I said. "I wrote my mother asking her to knit me a pair of gloves."

There is still something wrong with the wireless. We have received only one message in two weeks. It said, "Russian opposition collapsing," and there was so much static we could not make out the rest of it. We have taken the wireless apart twice. The first time we found a loose wire but the second time we could not find anything. If Hans were here he would be able to find the trouble immediately.

"I have a theory about the wireless," Muller says. He has had ten theories in as many days: The magnet of the Russians is pulling our signals in to it; the northern lights, which have been shifting uneasily on the horizon, make a curtain the wireless signals cannot get through; the Russian opposition is not collapsing at all. They are drawing us deeper and deeper into a trap.

I say, "I am going to try again. Perhaps the trouble has cleared up," and put the headphones on so I do not have to listen to his new theory. I can hear nothing but a rumbling roar that sounds like the front.

I take out the folded piece of paper Dr. Funkenheld gave me and lay it on the wireless. He comes nearly every night to see if I have gotten an answer to his message, and I take off the headphones and let him listen to the static. I tell him that we cannot get through, but even though that is true, it is not the real reason I have not sent the message. I am afraid of the commandant finding out. I am afraid of being sent to the front.

I have compromised by writing a letter to the professor that I studied medicine with in Jena, but I have not gotten an answer from him yet, and so I must go on pretending to the doctor.

"You don't have to do that," Muller says. He sits on the wireless, swinging his leg. He picks up the paper with the symptoms on it and holds it to the flame of the Primus stove. I grab for it, but it is already burning redly. "I have sent the message for you."

"I don't believe you. Nothing has been getting out."

"Didn't you notice the northern lights did not appear last night?"

I have not noticed. The ointment the doctor gave to me makes everything look red at night, and I do not believe in Muller's theories. "Nothing is getting out now," I say, and hold the headphones out to him so he can hear the static. He listens, still swinging his leg. "You will get us both in trouble. Why did you do it?"

"I was curious about it." If we are sent up to the front, his curiosity will kill us. He will take apart a land mine to see how it works. "We cannot get in trouble for sending military messages. I said the commandant was afraid it was a poisonous gas the Russians were using." He swings his leg and grins because now I am the curious one.

"Well, did you get an answer?"

"Yes," he says maddeningly and puts the headphones on. "It is not a poisonous gas."

I shrug as if I do not care whether I get an answer or not. I put on my cap and the muffler my mother knitted for me and open the door, "I am going out to see if the mail has come. Perhaps there will be a letter there from my professor."

"Nature of disease unknown," Muller shouts against the sudden force of the snow. "Possibly impetigo or glandular disorder."

I grin back at him and say, "If there is a package from my mother I will give you half of what is in it."

"Even if it is your gloves?"

"No, not if it is my gloves," I say, and go to find the doctor.

At the dressing station they tell me he has gone to see Schwarzschild and give me directions to the artillery staff's headquarters. It is not very far, but it is snowing and my hands are already cold. I go to the quartermaster's and ask him if the mail has come in.

There is a new recruit there, trying to fix Eisner's motorcycle. He has parts spread out on the ground all around him in a circle. He points to a burlap sack and says, "That is all the mail there is. Look through it yourself."

Snow has gotten into the sack and melted. The ink on the envelopes has run, and I squint at them, trying to make out the names. My eyes begin to hurt. There is not a package from my mother or a letter from my professor, but there is a letter for Lieutenant Schwarzschild. The return address says *Doctor*. Perhaps he has written to a doctor himself.

"I am delivering a message to the artillery headquarters," I say, showing the letter to the recruit. "I will take this up, too." The recruit nods and goes on working.

It has gotten dark while I was inside, and it is snowing harder. I jam my hands in the ice-stiff pockets of my coat and start to the artillery headquarters in the rear. It is pitch-dark in the communication trenches, and the wind twists the snow and funnels it howling along them. I take off my muffler and wrap it around my hands like a girl's muff.

A band of red shifts uneasily all along the horizon, but I do not know if it is the front or Muller's northern lights, and there is no shelling to guide me. We are running out of shells, so we do not usually begin shelling until nine o'clock. The Russians start even later. Sometimes I hear machine-gun fire, but it is distorted by the wind and the snow, and I cannot tell what direction it is coming from.

The communication trench seems narrower and deeper than

I remember it from when Hans and I first brought the wireless up. It takes me longer than I think it should to get to the branching that will lead north to the headquarters. The front has been contracting, the ammunition dumps and officer's billets and clearing stations moving up closer and closer behind us. The artillery headquarters has been moved up from the village to a dugout near the artillery line, not half a mile behind us. The nightly firing is starting. I hear a low rumble, like thunder.

The roar seems to be ahead of me, and I stop and look around, wondering if I can have gotten somehow turned around, though I have not left the trenches. I start again, and almost immediately I see the branching and the headquarters.

It has no door, only a blanket across the opening, and I pull my hands free of the muffler and duck through it into a tiny space like a rabbit hole, the timber balks of the earthen ceiling so low I have to stoop. Now that I am out of the roar of the snow, the sound of the front separates itself into the individual crack of a four-pounder, the whine of a star shell, and under it the almost continuous rattle of machine guns. The trenches must not be as deep here. Muller and I can hardly hear the front at all in our wireless hut.

A man is sitting at an uneven table spread with papers and books. There is a candle on the table with a red glass chimney, or perhaps it only looks that way to me. Everything in the dugout, even the man, looks faintly red. He is wearing a uniform but no coat, and gloves with the finger ends cut off, even though there is no stove here. My hands are already cold.

A trench mortar roars, and clods of frozen dirt clatter from the roof onto the table. The man brushes the dirt from the papers and looks up.

"I am looking for Dr. Funkenheld," I say.

"He is not here." He stands up and comes around the table, moving stiffly, like an old man, though he does not look older than forty. He has a moustache, and his face looks dirty in the red light.

"I have a message for him."

An eight-pounder roars, and more dirt falls on us. The man

raises his arm to brush the dirt off his shoulder. The sleeve of his uniform has been slit into ribbons. All along the back of his raised hand and the side of his arm are red sores running with pus. I look back at his face. The sores in his moustache and around his nose and mouth have dried and are covered with a crust. Excoriated lesions. Suppurating bullae. The gun roars again, and dirt rains down on his raw hands.

"I have a message for him," I say, backing away from him. I reach in the pocket of my coat to show him the message, but I pull out the letter instead. "There was a letter for you, Lieutenant Schwarzschild." I hold it out to him by one corner so he will not touch me when he takes it.

He comes toward me to take the letter, the muscles in his jaw tightening, and I think in horror that the sores must be on his legs as well. "Who is it from?" he says. "Ah, Herr Professor Einstein. Good," and turns it over. He puts his fingers on the flap to open the letter, and cries out in pain. He drops the letter.

"Would you read it to me?" he says and sinks down into the chair, cradling his hand against his chest. I can see there are sores in his fingernails.

I do not have any feeling in my hands. I pick the envelope up by its corners and turn it over. The skin of his finger is still on the flap. I back away from the table. "I must find the doctor. It is an emergency."

"You would not be able to find him," he says. Blood oozes out of the tip of his finger and down over the blister in his fingernail. "He has gone up to the front."

"What?" I say, backing and backing until I run into the blanket. "I cannot understand you."

"He has gone up to the front," he says, more slowly, and this time I can puzzle out the words, but they make no sense. How can the doctor be at the front? This is the front.

He pushes the candle toward me. "I order you to read me the letter."

I do not have any feeling in my fingers. I open it from the top, tearing the letter almost in two. It is a long letter, full of equations and numbers, but the words are warped and blurred.

" 'My Esteemed Colleague! I have read your paper with the greatest interest. I had not expected that one could formulate the exact solution of the problem so simply. The analytical treatment of the problem appears to me splendid. Next Thursday I will present the work with several explanatory words, to the Academy!' "

"Formulated so simply," Schwarzschild says, as if he is in pain. "That is enough. Put the letter down. I will read the rest of it."

I lay the letter on the table in front of him, and then I am running down the trench in the dark with the sound of the front all around me, roaring and shaking the ground. At the first turning, Muller grabs my arm and stops me. "What are you doing here?" I shout. "Go back! Go back!"

"Go back?" he says. "The front's that way." He points in the direction he came from. But the front is not that way. It is behind me, in the artillery headquarters. "I told you there would be a bombardment tonight. Did you see the doctor? Did you give him the message? What did he say?"

"So you actually held the letter from Einstein?" Travers said. "How exciting that must have been! Only two months after Einstein had published his theory of general relativity. And years before they realized black holes really existed. When was this exactly?" He took out a notebook and began to scribble notes. "My esteemed colleague . . ." he muttered to himself. "Formulated so simply. This is great stuff. I mean, I've been trying to find out stuff on Schwarzschild for my paper for months, but there's hardly any information on him. I guess because of the war."

"No information can get out of a black hole once the Schwarzschild radius has been passed," I said.

"Hey, that's great!" he said, scribbling. "Can I use that in my paper?"

Now I am the one who sits endlessly in front of the wireless sending out messages to the Red Cross, to my professor in Jena, to Dr. Einstein. I have frostbitten the forefinger and thumb of my right hand and have to tap out the letters with my left. But nothing is getting out, and I must get a message out. I must find someone to tell me the name of Schwarzschild's disease.

"I have a theory," Muller says. "The Jews have seized power and have signed a treaty with the Russians. We are completely cut off."

"I am going to see if the mail has come," I say, so that I do not have to listen to any more of his theories, but the doctor stops me on my way out the hut.

I tell him what the message said. "Impetigo!" the doctor shouts. "You saw him! Did that look like impetigo to you?"

I shake my head, unable to tell him what I think it looks like.

"What are his symptoms?" Muller asks, burning with curiosity. I have not told him about Schwarzschild. I am afraid that if I tell him, he will only become more curious and will insist on going up to the front to see Schwarzschild himself.

"Let me see your eyes," the doctor says in his beautiful calm voice. I wish he would ask Muller to go for a hand lamp again so that I could ask him how Schwarzschild is, but he has brought a candle with him. He holds it so close to my face that I cannot see anything but the red flame.

"Is Lieutenant Schwarzschild worse? What are his symptoms?" Muller says, leaning forward.

His symptoms are craters and shell holes, I think. I am sorry I have not told Muller, for it has only made him more curious. Until now I have told him everything, even how Hans died when the wireless hut was hit, how he laid the liquid barretter carefully down on top of the wireless before he tried to cough up what was left of his chest and catch it in his hands. But I cannot tell him this.

"What symptoms does he have?" Muller says again, his nose almost in the candle's flame, but the doctor turns from him as if he cannot hear him and blows the candle out. The doctor unwraps the dressing and looks at my fingers. They are swollen and red. Muller leans over the doctor's shoulder. "I have a theory about Lieutenant Schwarzschild's disease," he says.

"Shut up," I say. "I don't want to hear any more of your stupid theories," and do not even care about the wounded look on Muller's face or the way he goes and sits by the wireless. For now I have a theory, and it is more horrible than anything Muller could have dreamt of.

We are all of us—Muller, and the recruit who is trying to put together Eisner's motorcycle, and perhaps even the doctor with his steady bedside voice—afraid of the front. But our fear is not complete, because unspoken in it is our belief that the front is something separate from us, something we can keep away from by keeping the wireless or the motorcycle fixed, something we can survive by flattening our faces into the frozen earth, something we can escape altogether by being invalided out.

But the front is not separate. It is inside Schwarzschild, and the symptoms I have been sending out, suppurative bullae and excoriated lesions, are not what is wrong with him at all. The lesions on his skin are only the barbed wire and shell holes and connecting trenches of a front that is somewhere farther in.

The doctor puts a new dressing of crepe paper on my hand. "I have tried to invalid Schwarzschild out," the doctor says, and Muller looks at him, astounded. "The supply lines are blocked with snow."

"Schwarzschild cannot be invalided out," I say. "The front is inside him."

The doctor puts the roll of crepe paper back in his kit and closes it. "When the roads open again, I will invalid you out for frostbite. And Muller too."

Muller is so surprised he blurts, "I do not have frostbite."

But the doctor is no longer listening. "You must both escape," he says—and I am not sure he is even listening to himself—"while you can."

"I have a theory about why you have not told me what is wrong with Schwarzschild," Muller says as soon as the doctor is gone.

"I am going for the mail."

"There will not be any mail," Muller shouts after me. "The supply lines are blocked," but the mail is there, scattered among the motorcycle parts. There are only a few parts left. As soon as the roads are cleared, the recruit will be able to climb on the motorcycle and ride away.

I gather up the letters and take them over to the lantern to try to read them, but my eyes are so bad I cannot see anything but a red blur. "I am taking them back to the wireless hut," I say, and the recruit nods without looking up.

It is starting to snow. Muller meets me at the door, but I brush past him and turn the flame of the Primus stove up as high as it will go and hold the letters up behind it.

"I will read them for you," Muller says eagerly, looking through the envelopes I have discarded. "Look, here is a letter from your mother. Perhaps she has sent your gloves."

I squint at the letters one by one while he tears open my mother's letter to me. Even though I hold them so close to the flame that the paper scorches, I cannot make out the names.

" 'Dear son,' " Muller reads, 'I have not heard from you in three months. Are you hurt? Are you ill? Do you need anything?' "

The last letter is from Professor Zuschauer in Jena. I can see his name quite clearly in the corner of the envelope, though mine is blurred beyond recognition. I tear it open. There is nothing written on the red paper.

I thrust it at Muller. "Read this," I say.

"I have not finished with your mother's letter yet," Muller says, but he takes the letter and reads: " 'Dear Herr Rottschieben, I received your letter yesterday. I could hardly decipher your writing. Do you not have decent pens at the front? The disease you describe is called Neumann's disease or pemphigus—' "

I snatch the letter out of Muller's hands and run out the door. "Let me come with you!" Muller shouts.

"You must stay and watch the wireless!" I say joyously, running along the communication trench. Schwarzschild does not have the front inside him. He has pemphigus, he has Neumann's disease, and now he can be invalided home to hospital.

I go down and think I have tripped over a discarded helmet

or a tin of beef, but there is a crash, and dirt and revetting fall all around me. I hear the low buzz of a daisy cutter and flatten myself into the trench, but the buzz does not become a whine. It stops, and there is another crash and the trench caves in.

I scramble out of the trench before it can suffocate me and crawl along the edge toward Schwarzschild's dugout, but the trench has caved in all along its length, and when I crawl up and over the loose dirt, I lose it in the swirling snow.

I cannot tell which way the front lies, but I know it is very close. The sound comes at me from all directions, a deafening roar in which no individual sounds can be distinguished. The snow is so thick I cannot see the burst of flame from the muzzles as the guns fire, and no part of the horizon looks redder than any other. It is all red, even the snow.

I crawl in what I think is the direction of the trench, but as soon as I do, I am in barbed wire. I stop, breathing hard, my face and hands pressed into the snow. I have come the wrong way. I am at the front. I hear a sound out of the barrage of sound, the sound of tires on the snow, and I think it is a tank, and cannot breathe at all. The sound comes closer, and in spite of myself I look up and it is the recruit who was at the quartermaster's.

He is a long way away, behind a coiled line of barbed wire, but I can see him quite clearly in spite of the snow. He has the motorcycle fixed, and as I watch, he flings his leg over it and presses his foot down. "Go!" I shout. "Get out!" The motorcycle jumps forward. "Go!"

The motorcycle comes toward me, picking up speed. It rears up, and I think it is going to jump the barbed wire, but it falls instead, the motorcycle first and then the recruit, spiraling slowly down into the iron spikes. The ground heaves, and I fall too.

I have fallen into Schwarzschild's dugout. Half of it has caved in, the timber balks sticking out at angles from the heap of dirt and snow, but the blanket is still over the door, and Schwarzschild is propped in a chair. The doctor is bending over him. Schwarzschild has his shirt off. His chest looks like Hans's did.

The front roars and more of the roof crumbles. "It's all right!

It's a disease!" I shout over it. "I have brought you a letter to prove it," and hand him the letter which I have been clutching in my unfeeling hand.

The doctor grabs the letter from me. Snow whirls down through the ruined roof, but Schwarzschild does not put on his shirt. He watches uninterestedly as the doctor reads the letter.

" 'The symptoms you describe are almost certainly those of Neumann's disease, or pemphigus vulgaris. I have treated two patients with the disease, both Jews. It is a disease of the mucous membranes and is not contagious. Its cause is unknown. It always ends in death.' " Dr. Funkenheld crumples up the paper. "You came all this way in the middle of a bombardment to tell me there is no hope?" he shouts in a voice I do not even recognize, it is so unlike his steady doctor's voice. 'You should have tried to get away. You should have—" and then he is gone under a crashing of dirt and splintered timbers.

I struggle toward Schwarzschild through the maelstrom of red dust and snow. "Put your shirt on!" I shout at him. "We must get out of here!" I crawl to the door to see if we can get out through the communication trench.

Muller bursts through blanket. He is carrying, impossibly, the wireless. The headphones trail behind him in the snow. "I came to see what had happened to you. I thought you were dead. The communication trenches are shot to pieces."

It is as I had feared. His curiosity has got the best of him, and now he is trapped, too, though he seems not to know it. He hoists the wireless onto the table without looking at it. His eyes are on Schwarzschild, who leans against the remaining wall of the dugout, his shirt in his hands.

"Your shirt!" I shout and come around to help Schwarzschild put it on over the craters and shell holes of his blasted skin. The air screams and the mouth of the dugout blows in. I grab at Schwarzschild's arm, and the skin of it comes off in my hands. He falls against the table, and the wireless goes over. I can hear the splintering tinkle of the liquid barretter breaking, and then the whole dugout is caving in and we are under the table. I cannot see anything.

"Muller!" I shout. "Where are you?"

"I'm hit," he says.

I try to find him in the darkness, but I am crushed against Schwarzschild. I cannot move. "Where are you hit?"

"In the arm," he says, and I hear him try to move it. The movement dislodges more dirt, and it falls around us, shutting out all sound of the front. I can hear the creak of wood as the table legs give way.

"Schwarzschild?" I say. He doesn't answer, but I know he is not dead. His body is as hot as the Primus stove flame. My hand is underneath his body, and I try to shift it, but I cannot. The dirt falls like snow, piling up around us. The darkness is red for a while, and then I cannot see even that.

"I have a theory," Muller says in a voice so close and so devoid of curiosity it might be mine. "It is the end of the world."

"Was that when Schwarzschild was sent home on sick leave?" Travers said.

"Or validated, or whatever you Germans call it? Well, yeah, it had to be, because he died in March. What happened to Muller?"

I had hoped he would go away as soon as I had told him what had happened to Schwarzschild, but he made no move to get up. "Muller was invalided out with a broken arm. He became a scientist."

"The way you did." He opened his notebook again. "Did you see Schwarzschild after that?"

The question makes no sense.

"After you got out? Before he died?"

It seems to take a long time for his words to get to me. The message bends and curves, shifting into the red, and I can hardly make it out. "No," I say, though that is a lie.

Travers scribbles. "I really do appreciate this, Dr. Rottschieben. I've always been curious about Schwarzschild, and now that you've told me all this stuff I'm even more interested," Travers says, or seems to say. Messages coming in are warped by the gravitational blizzard into something that no longer resembles speech. "If you'd be willing to help me, I'd like to write my thesis on him."

Go. Get out. "It was a lie," I say. "I never knew Schwarzschild. I saw him once, from a distance—your fixed observer."

Travers looks up expectantly from his notes as if he is still waiting for me to answer him.

"Schwarzschild was never even in Russia," I lie. "He spent the whole winter in hospital in Göttingen. I lied to you. It was nothing but a thought problem."

He waits, pencil ready.

"You can't stay here!" I shout. "You have to get away. There is no safe distance from which a fixed observer can watch without being drawn in, and once you are inside the Schwarzschild radius you can't get out. Don't you understand? We are still there!"

We are still there, trapped in the trenches of the Russian front, while the dying star burns itself out, spiraling down into that center where time ceases to exist, where everything ceases to exist except the naked singularity that is somehow Schwarzschild.

Muller tries to dig the wireless out with his crushed arm so he can send a message that nobody can hear—"Help us! Help us!"—and I struggle to free the hands that in spite of Schwarzschild's warmth are now so cold I cannot feel them, and in the very center Schwarzschild burns himself out, the black hole at his center imploding him cell by cell, carrying him down into darkness, and us with him.

"It is a trap!" I shout at Travers from the center, and the message struggles to escape and then falls back.

"I wonder how he figured it out?" Travers says, and now I can hear him clearly. "I mean, can you imagine trying to figure out something like the theory of black holes in the middle of a war and while you were suffering from a fatal disease? And just think, when he came up with the theory, he didn't have any idea that black holes even existed."

Letter from Caroline Herschel (1750–1848)

Siv Cedering

William is away, and I am minding
the heavens. I have discovered
eight new comets and three nebulae
never before seen by man,
and I am preparing an Index to
Flamsteed's observations, together with
a catalogue of 560 stars omitted from
the British Catalogue, plus a list of errata
in that publication. William says

I have a way with numbers, so I handle
all the necessary reductions and
calculations. I also plan
every night's observation
schedule, for he says my intuition
helps me turn the telescope to discover
star cluster after star cluster.

I have helped him polish the mirrors
and lenses of our new telescope. It is
the largest in existence. Can you imagine
the thrill of turning it to some new
corner of the heavens to see

something never before seen
from earth? I actually like

that he is busy with the Royal society
and his club, for when I finish my other work
I can spend all night sweeping
the heavens.

Sometimes when I am alone
in the dark, and the universe reveals
yet another secret, I say the names
of my long, lost sisters, forgotten
in the books that record
our science—

> Aganice of Thessaly,
> Hyptia,
> Hildegard,
> Catherina Hevelius,
> Maria Agnesi

—as if the stars themselves could

remember. Did you know that Hildegard
proposed a heliocentric universe
300 years before Copernicus? that she
wrote of universal gravitation 500 years
before Newton? But who would listen
to her? She was just a nun, a woman.
What is our age, if that age was dark?
As for my name, it will also be
forgotten, but I am not accused
of being a sorceress, like Aganice,
and the Christians do not threaten to
drag me to church, to murder me, like they did
Hyptia of Alexandria, the eloquent, young
woman who devised the instruments
used to accurately measure the position

and motion of
heavenly bodies.
However long we live, life is short, so I
work. And however important man becomes,
he is nothing compared to the stars.
There are secrets, dear sister, and it is
for us to reveal them. Your name, like mine,
is a song. Write soon,

Caroline

From *We*

Yevgeny Zamyatin

Translated by Gregory Zilboorg

Though written in Russian and completed in 1921, Yevgeny Zamyatin's great dystopian novel *We* was first published in English, by the New York publisher E. P. Dutton, in 1924. The complete Russian text was not published until 1952, again in New York, and the book could not be issued in its author's native land until 1988.

Set several centuries in the future, *We* depicts the journal of D-503, a mathematician and leading citizen of OneState, and builder of its first spacecraft, the *Integral*. At first D-503 intends his journal to record the achievements of a perfect society, "a derivative of our life, of the mathematically perfect life of OneState." But he meets a woman, I-330, whose extraordinary sense of personal freedom repels and intrigues him. The first two entries in this selection occur shortly after he has met I-330. By the time of the second pair of entries, D-503 and I-330 have had several illicit meetings, he has fallen in love with her and also begun to suspect that she belongs to a subversive organization intent on disrupting the seamless life of OneState. D-503 is suffering irrational passions, which he attributes to illness. It turns out that there is something of an epidemic.

Record 4

Savage with Barometer
Epilepsy
If

Up to this point I have found everything in life clear (not for nothing do I seem to have a certain partiality for that very word *clear*). But today . . . I don't understand.

First: I did in fact get an order to be in that very auditorium 112, just as she had told me. Although the probability was something like:

$$\frac{1500}{10{,}000{,}000} = \frac{3}{20{,}000}$$

1500 is the number of auditoriums, and 10,000,000 the number of Numbers.

Second: It might be best, however, to go in order.

The auditorium. An immense sunlit hemisphere composed of massive glass sections. Circular rows of nobly spherical, smoothly shaved heads. I looked around with a slightly sinking heart. I think I was searching whether the pink crescent of my dear O's lips would not shine above the blue waves of the yunies. There . . . it looked like someone's very white, shiny teeth . . . but no, not hers. This evening at 21:00 hours O was to come to my place—it was perfectly natural that I'd want to see her here.

The bell. We stood up and sang the Anthem of OneState, and on the platform appeared the phonolecturer, sparkling with wit and with his golden loudspeaker.

"Honored Numbers! Recently our archaeologists unearthed a book of the twentieth century. The author ironically relates the story of the savage and the barometer. The savage noticed that every time the barometer showed *Rain,* it did in fact rain. And, since the savage wanted it to rain, he found a way to let out just enough mercury so that the thing would point to *Rain.* [The screen showed a savage bedecked with feathers letting out some mercury: general laughter.] You laugh. But don't you think the

European of that age was much more to be laughed at? Just like the savage, the European also wanted 'rain.' But *Rain* with a capital letter, algebraic *Rain*. But he stood in front of the barometer like a wet hen. The savage, at least, had more daring, more energy, and more—even if savage—logic. He was able to establish a connection between a cause and its effect. When he let that mercury out, he took the first step along the great path that . . ."

But at this (I repeat: I'm writing what happened, leaving nothing out) I became for a time impermeable to the vivifying stream pouring out of the loudspeaker. It suddently struck me that I shouldn't have come (why "shouldn't," and how could I not have come, once I'd gotten the order?); it suddenly struck me that everything was empty, an empty shell. And I didn't manage to switch my attention back on until the phonolecturer had gotten down to his basic theme: to our music, to mathematical composition (the mathematician is the cause, the music the result), to a description of the recently invented musicometer.

". . . Simply by turning this handle, any one of you can produce up to three sonatas per hour. And how much labor such a thing cost your ancestors! They could create only by whipping themselves up to attacks of 'inspiration'—some unknown form of epilepsy. And here I have for you a most amusing example of what they got for their trouble—the music of Scriabin, twentieth century. This black box (a curtain was pulled aside on the stage, and there stood one of their ancient instruments), this black box was called a 'grand piano,' or even a 'Royal Grand,' which is merely one more proof, if any were needed, of the degree to which all their music . . ."

And then . . . but again I'm not sure, because it might have been . . . no, I'll say it right out . . . because she, I-330, went up to the "Royal Grand." I was probably just dazzled by how she suddenly turned up, unexpectedly, on the stage.

She was wearing one of the fantastic costumes of ancient times: a tightly fitting black dress, very low cut, which sharply emphasized the whiteness of her shoulders and bosom, and the warm shadow that undulated in time with her breathing between her . . . and her blinding, almost wicked teeth. . . .

Her smile was a bite, and I was its target. She sat down. She began to play. Something wild, spasmodic, jumbled—like their whole life back then, when they didn't have even the faintest adumbration of rational mechanics. And of course those around me were right to laugh, as they all did. But a few of us . . . and I . . . why was I among those few?

Yes, epilepsy is a mental illness—pain . . . A slow, sweet pain—it is a bite—let it bite deeper, harder. And then, slowly, the sun. Not this one, not ours, shining all sky-blue crystal regularity through the glass brick—no: a savage, rushing, burning sun—flinging everything away from itself—everything in little pieces.

The one sitting next to me glanced to his left, at me, and giggled. For some reason I have a very vivid image of what I saw: a microscopic bubble of saliva appeared on his lips and burst. That bubble sobered me. I was myself once again.

Like everyone else, I heard nothing more than the stupid vain clattering of the strings. I laughed. Things became easy and simple. The talented phonolecturer had simply given us a too lively picture of that savage epoch—that's all.

After that, how pleasant it was to listen to our music of today. (A demonstration of it was given at the end, for contrast.) Crystalline chromatic scales of converging and diverging infinite series—and the synoptic harmonies of the formulae of Taylor and Maclaurin, wholesome, quadrangular, and weighty as Pythagoras's pants; mournful melodies of a wavering, diminishing movement, the alternating bright beats of the pauses according to the lines of Frauenhofer—the spectral analysis of the planet . . . What magnificence! What unalterable regularity! And what pathetic self-indulgence was that ancient music, limited only by its wild imaginings. . . .

We left through the broad doors of the auditorium in the usual way, marching four abreast in neat ranks. I caught a glimpse of the familiar double-bent figure somewhere to the side and bowed to him respectfully.

Dear O was to come in an hour. I felt a pleasant and useful excitement. Once home I passed quickly by the desk, handed the duty officer my pink ticket, and got the pass to use the blinds. We get to use the blinds only on Sex Day. Otherwise we live in broad daylight inside these walls that seem to have been fashioned out of bright air, always on view. We have nothing to hide from one another. Besides, this makes it easier for the Guardians to carry out their burdensome, noble task. No telling what might go on otherwise. Maybe it was the strange opaque dwellings of the ancients that gave rise to their pitiful cellular psychology. "My [*sic*] home is my castle!" Brilliant, right?

At 22:00 hours I lowered the blinds—and at that precise moment O came in, a little out of breath. She gave me her pink lips—and her pink ticket. I tore off the stub, but I couldn't tear myself away from her rosy lips until the very last second: 22:15.

Afterward I showed her my "notes" and spoke—rather well, I think—about the beauty of the square, the cube, the straight line. She listened in her enchantingly rosy way. . . . and suddenly a tear fell from her blue eyes . . . then a second, a third . . . right on the page that was open (page 7). Made the ink run. So . . . I'll have to copy it over.

"Dear Dee, if only you . . . if . . ."

Well, what does that "if" mean? "If" what? She was singing the same old tune again: a child. Or maybe it was something new . . . about . . . about that other one. Though even here it seemed as though . . . But no, that would be too stupid.

Record 5

Square
Rulers of the World
Pleasant and Useful Function

Wrong again. Again I'm talking to you, my unknown reader, as though you were . . . well, say, as though you were my old comrade R-13, the poet, the one with the African lips—everyone knows him. You, meanwhile, you might be anywhere . . . on the moon, on Venus, on Mars, on Mercury. Who knows you, where you are and who you are?

Here's what: Imagine a square, a splendid, living square. And he has to tell about himself, about his life. You see—the last thing on earth a square would think of telling about is that he has four equal angles. He simply does not see that, it's so familiar to him, such an everyday thing. That's me, I'm in the situation of that square all the time. Take the pink tickets and all that—to me that's nothing more than the four equal angles, but for you that might be, I don't know, as tough as Newton's binomial theorem.

So. One of the ancient wise men—by accident, of course—managed to say something very smart: "Love and hunger rule the world." *Ergo:* To rule the world, man has got to rule the rulers of the world. Our forebears finally managed to conquer Hunger, by paying a terrible price: I'm talking about the 200-Years War, the war between the City and the Country. It was probably religious prejudice that made the Christian savages fight so stubbornly for their "bread."* But in the year 35 before the founding of OneState our present petroleum food was invented. True, only 0.2 of the world's population survived. On the other hand, when it was cleansed of a thousand years of filth, how bright the face of the earth became! And what is more, the zero point two tenths who survived . . . tasted earthly bliss in the granaries of OneState.

But isn't it clear that bliss and envy are the numerator and denominator of that fraction known as happiness? And what sense would there be in all the numberless victims of the 200-Years War if there still remained in our life some cause for envy? But some cause did remain, because noses remained, the button noses and classical noses mentioned in that conversation on our walk, and because there are some whose love many people want, and others whose love nobody wants.

It's natural that once Hunger had been vanquished (which is algebraically the equivalent of attaining the summit of material well-being), OneState mounted an attack on that other ruler of the world, Love. Finally, this element was also conquered, i.e., organized, mathematicized, and our *Lex sexualis* was promulgated

* This word has come down to us only as a poetic metaphor. It is not known what the chemical composition of this material was.

about 300 years ago: "Any Number has the right of access to any other Number as sexual product."

The rest is a purely technical matter. They give you a careful going-over in the Sexual Bureau labs and determine the exact content of the sexual hormones in your blood and work out your correct Table of Sex Days. Then you fill out a declaration that on your days you'd like to make use of Number (or Numbers) so-and-so and they hand you the corresponding book of tickets (pink). And that's it.

So it's clear—there's no longer the slightest cause for envy. The denominator of the happiness fraction has been reduced to zero and the fraction becomes magnificent infinity. And the very same thing that the ancients found to be a source of endless tragedy became for us a harmonious, pleasant, and useful function of the organism, just like sleep, physical work, eating, defecating, and so on. From this you can see how the mighty power of logic cleanses whatever it touches. Oh, if only you, my unknown readers, could come to know this divine power, if only you too could follow it to the end!

Strange—today I've been writing about the loftiest summits of human history, the whole time I've been breathing the purest mountain air of thought . . . but inside there is something cloudy, something spidery, something cross-shaped like that four-pawed X. Or is it my own paws bothering me, the fact that they've been in front of my eyes so long, these shaggy paws? I don't like talking about them. I don't like them. They're a holdover from the savage era. Can it really be true that I contain . . .

I wanted to cross all that out . . . because that's beyond the scope of these notes. But then I decided: No, I'll leave it in. Let these notes act like the most delicate seismograph, let them register the least little wiggles in my brainwaves, however insignificant. Sometimes, you never know, these are just the wiggles that give you the first warning . . .

But that's absurd, now. I really should cross it out. We've channeled all the elements of nature. No catastrophe can happen.

But now it's perfectly clear to me: That strange inner feeling just comes from my being like the square I spoke about earlier.

And there's no X in me (there could not be)—I'm simply worried that there might still be some X in you, my unknown readers. But I have faith that you won't judge me too harshly.

I have faith that you will understand how hard it is for me to write, harder than for any other writer in the whole extent of human history: They wrote for their contemporaries, others wrote for posterity, but nobody ever wrote for their ancestors or for people like their wild remote ancestors. . . .

RECORD 20

Discharge
Idea Material
Zero Cliff

Discharge is the most suitable definition. That's what that was, I now see: like an electrical discharge. These last few days my pulse has been getting drier and drier, quicker and quicker, more and more tense—the poles closer and closer, making this dry crackling sound. One millimeter more, and there'll be an explosion. After which: silence.

Inside me right now it's very quiet and empty, the same as in the building when everyone's left and you're lying all alone, sick, and you can hear this clear, precise, metallic beating of your thoughts.

Maybe this "discharge" cured me, finally, of that torment called my "soul," and I'm just like all the rest of us again. Now at least I don't feel any pain when I see O in my thoughts standing on the steps of the Cube, when I see her under the Gas Bell. And if she gives them my name there in Operations—so be it. My last act will be to put a pious and grateful kiss on the Benefactor's punishing hand. In my relationship with OneState I have that right, to undergo punishment, and this right I will not give up. None of us Numbers ought, none dares, to refuse that one right that we have, which means it is the most valuable.

. . . There's a quiet, clear metallic sound to my thoughts' clicking; an unknown aero is carrying me away into the blue heights of my favorite abstractions. And here in this purest, most rarefied air, I see my reflection "on operative right" pop with a slight bang like a tire blowing out. And I see clearly that it was nothing more than a throwback to the ancients' idiotic superstition, their idea about one's "right."

There are ideas made of clay, and there are ideas sculpted for the ages out of gold or out of our precious glass. And to determine what material an idea is made of, all you have to do is let a drop of powerful acid fall on it. Even the ancients knew one such acid: *reductio ad finem*. That's what they seem to have called it. But they were afraid of this poison. They preferred to see at least some kind of heaven—however clay, however toylike—to this blue nothing. But we are grown-ups, thanks be to the Benefactor, and don't need toys.

Look here—suppose you let a drop fall on the idea of "rights." Even among the ancients the more grown-up knew that the source of right is power, that right is a function of power. So, take some scales and put on one side a gram, on the other a ton; on one side "I" and on the other "We," OneState. It's clear, isn't it?—to assert that "I" has certain "rights" with respect to the State is exactly the same as asserting that a gram weighs the same as a ton. That explains the way things are divided up: To the ton go the rights, to the gram the duties. And the natural path from nullity to greatness is this: Forget that you're a gram and feel yourself a millionth part of a ton.

You, plump, rosy-cheeked Venusians, and you Uranites, sooty as blacksmiths—in my blue silence I can hear your grumbling. But understand this: All greatness is simple. Understand this: Only the four rules of arithmetic are unalterable and everlasting. And only that moral system built on the four rules will prevail as great, unalterable, and everlasting. That is the ultimate wisdom. That is the summit of the pyramid up which people, red and sweating, kicking and panting, have scrambled for centuries. And looking down from this summit to the bottom, we see what remains in us of our savage ancestors seething like wretched worms. Looking down from this summit, there's no difference between a woman who gave birth illegally—O—and a murderer, and that madman who dared aim his poem at OneState. And the verdict is the same for them all: premature death. This is the very same divine justice dreamt of by the people of the stone-house age, illuminated by the rosy naive rays of the dawn of history: Their "God" punished abuse of Holy Church exactly the same as murder.

You Uranites, stern and black as early Spaniards, you who were wise enough to do some burning at the stake, you are silent; I think you're with me. But you rosy Venusians . . . among you I hear something about torture, executions, a return to the age of barbarism. I'm sorry for you, old dears. You aren't up to philosophical-mathematical thinking.

Human history ascends in spirals, like an aero. The circles vary, some are gold, some are bloody, but all are divided into the same 360 degrees. It starts at zero and goes forward: 10, 20, 200, 360 degrees—then back to zero. Yes, we've come back to zero—yes. But for my mind, thinking in mathematics as it does, one thing is clear: This zero is completely different, new. Leaving zero, we headed to the right. We returned to zero from the left. So instead of plus zero, we have minus zero. Do you understand?

I see this zero as some kind of silent, huge, narrow, knife-sharp cliff. In ferocious, shaggy darkness, holding our breath, we pushed off from the black night side of the Zero Cliff. For centuries, like some new Columbus, we sailed and sailed and rounded the whole earth, and at last, Hurrah! A salute! All hands on deck and lookouts aloft! In front of us is the new, hitherto unknown side of the Zero Cliff, lit by the polar effulgence of OneState, a blue massif, the sparks of a rainbow, the sun, hundreds of suns, billions of rainbows. . . .

So what if nothing but the breadth of a knife blade separates us from the other dark side of the Zero Cliff? A knife is the most permanent, the most immortal, the most ingenious of all of man's creations. The knife was a guillotine, the knife is a universal means of resolving all knots, and the path of paradox lies along the blade of a knife—the only path worthy of the mind without fear. . . .

Record 22

Frozen Waves
Everything Tends to Perfection
I Am a Microbe

Imagine this: You're standing on a bank, and the waves rise up right on time, then, when they've crested, suddenly they stop, they freeze solid. That's just how terrifying and unnatural this was, too, when our walk, prescribed by the Tables, suddenly went haywire, this way and that, and stopped. According to our chronicles, the last time anything like this happened was 119 years ago when a meteorite came crashing out of the sky and landed, screaming and smoking, right in the midst of our walk.

We were walking the same as always, that is to say, just like the warriors you see on Assyrian monuments: a thousand heads with two fused, integrated legs, with two integrated arms, swinging wide. Down at the end of the avenue, where the Accumulator Tower was making its grim hum, a rectangle was coming toward us: The front, back, and two sides consisted of guards, and in the middle were three people with the gold numbers already gone from their yunies, and the whole thing was clear to the point of pain.

The huge clockface at the top of the tower—that was a face, leaning out of the clouds and spitting out the seconds and waiting. It couldn't care less. And then, right on the dot at 13:06, something crazy happened in the rectangle. It was all very close to me, I could see every smallest detail, and I very clearly remember a long thin neck and a tangle of twisted little light blue veins at the temples, like rivers on the map of some unknown little world, and this unknown world, it seemed, was a young man. He probably noticed someone in our ranks, got up on his tiptoes, craned his neck out, and stopped. One of the guards snapped him with the bluish spark of an electric knout. He gave out a thin squeal, like a puppy. This brought another neat snap, about one every two seconds—squeal, snap, squeal. . . .

We walked on as before in our measured, Assyrian walk, and I looked at the elegant zigzag of the sparks and thought: "Everything in human society is endlessly perfecting itself . . . and should perfect itself. What an ugly weapon the ancient knout was—and what beauty there is in . . ."

But at this point, like a nut flying off a machine at top speed, a lithe, slender female figure tore off out of our ranks screaming, "That's enough! Don't you dare . . . !" and pitched right into the midst of the rectangle. This was like the meteor 119 years ago. The whole walk froze, and our ranks were like the gray crests of waves instantly immobilized by a flash frost.

For one second I stared at her like all the others as something that had dropped out of nowhere: She was no longer a Number, she was simply a person; she existed as nothing more than the metaphysical substance of the insult committed against OneState. But then some one of her movements—turning, she twisted her hips to the left—and all at once I knew: I know her, I know that body resilient as a whip—my eyes, my lips, my hands know it—in one moment I was absolutely sure of it.

Two of the guards moved to cut her off. A still-bright mirror-like patch of pavement showed me that now, any moment, their trajectories would cross, they would grab her. My heart swallowed hard and stopped, and, never stopping to think whether one could or not, whether it was stupid or not, I dashed for that point. . . .

I could feel a thousand eyes popping with horror on me, but that only added more desperate giddy strength to the hairy-handed savage that had leapt out of me, and he ran all the faster. Two steps more, and she turned around. . . .

What I saw was a trembling face with lots of freckles and reddish eyebrows. It wasn't her. It wasn't I-330.

Wild, bursting joy. I feel like shouting something like, "Go on! Grab her!" but all I can hear coming out of me is a whisper. Then I feel a heavy hand take me by the shoulder, they're holding me, they're taking me somewhere, I'm trying to explain to them . . .

"Wait a minute, listen, you've got to understand, I thought she was . . ."

But how was I going to explain my whole being, this whole disease that I've been jotting down in these pages? So I shut up and went along peacefully. A leaf torn off a tree by a sudden gust of wind falls peacefully downward, but on the way it twists and catches at every twig, offshoot, branch that it knows. That is just how I caught at each silent, spherical head, at the transparent ice of the walls, at the light blue needle of the Accumulator Tower as it thrust up into the clouds.

At that moment, when a blank curtain was on the point of separating me forever from that whole splendid world, I see a little way off over the mirrored pavement a large familiar head and a pair of waving pink winglike arms. And I hear the familiar flat voice:

"I consider it my duty to state that Number D-503 is ill and in no condition to control his emotions. And I am convinced that he was carried away by a perfectly natural outrage. . . ."

"Yes, yes!" I jumped in. "I even shouted: Grab her!"

"You shouted nothing." This from just behind me.

"Yes, but I meant to . . . I swear by the Benefactor, I meant to."

I was instantly pierced by the cold gray eye-drills. I don't know whether he could see into me and tell that this was (almost) the truth, or whether he had some sort of secret aim to spare me once again for a while, but all he did was write out some note and hand it to one of the men holding me—and I was free once more, or rather, I was once more confined within the orderly, endless Assyrian ranks.

The rectangle containing the freckled face and the temple with the geographical map of blue veins disappeared round the corner, forever. We march on, one million-headed body, and each one of us harbors that humble happiness that also, probably, sustains the life of molecules, atoms, and phagocytes. In the ancient world, this was understood by the Christians, our only (if very imperfect) predecessors: Humility is a virtue, pride a vice; *We* comes from God, *I* from the Devil.

So here am I, in step with everyone else, and yet separate from all of them. I'm still trembling all over from the recent excitement—like a bridge that one of the ancient iron trains has just rumbled over. I feel myself. But it's only the eye with a lash in it, the swollen finger, the infected tooth that feels itself, is conscious of its own individual being. The healthy eye or finger or tooth doesn't seem to exist. So it's clear, isn't it? Self-consciousness is just a disease.

Maybe I'm no longer a phagocyte calmly going about the business of devouring microbes (with blue temples and freckles). Maybe I'm a microbe, and maybe there are thousands more of them among us, pretending like me to be phagocytes. . . .

This thing that happened today was really not so important, but just suppose it were only the beginning, only the first meteor of a whole shower of thundering hot rocks poured down by infinity onto our glass paradise?

The Garden of Time

J. G. Ballard

Towards evening, when the great shadow of the Palladian villa filled the terrace, Count Axel left his library and walked down the wide rococo steps among the time flowers. A tall, imperious figure in a black velvet jacket, a gold tie-pin glinting below his George V beard, cane held stiffly in a white-gloved hand, he surveyed the exquisite crystal flowers without emotion, listening to the sounds of his wife's harpsichord, as she played a Mozart rondo in the music room, echo and vibrate through the translucent petals.

The garden of the villa extended for some two hundred yards below the terrace, sloping down to a miniature lake spanned by a white bridge, a slender pavilion on the opposite bank. Axel rarely ventured as far as the lake; most of the time flowers grew in a small grove just below the terrace, sheltered by the high wall which encircled the estate. From the terrace he could see over the wall to the plain beyond, a continuous expanse of open ground that rolled in great swells to the horizon, where it rose slightly before finally dipping from sight. The plain surrounded the house on all sides, its drab emptiness emphasizing the seclusion and mellowed magnificence of the villa. Here, in the garden, the air seemed brighter, the sun warmer, while the plain was always dull and remote.

As was his custom before beginning his evening stroll, Count Axel looked out across the plain to the final rise, where the horizon was illuminated like a distant stage by the fading sun. As the Mozart chimed delicately around him, flowing from his wife's graceful hands, he saw that the advance column of an enormous army was moving slowly over the horizon. At first glance, the long ranks seemed to be progressing in orderly lines, but on closer inspection, it was apparent that, like the obscured detail of a Goya landscape, the army was composed of a vast throng of people, men and women, interspersed with a few soldiers in ragged uniforms, pressing forward in a disorganized tide. Some laboured under heavy loads suspended from crude yokes around their necks, others struggled with cumbersome wooden carts, their hands wrenching at the wheel spokes, a few trudged on alone, but all moved on at the same pace, bowed backs illuminated in the fleeting sun.

The advancing throng was almost too far away to be visible, but even as Axel watched, his expression aloof yet observant, it came perceptibly nearer, the vanguard of an immense rabble appearing from below the horizon. At last, as the daylight began to fade, the front edge of the throng reached the crest of the first swell below the horizon, and Axel turned from the terrace and walked down among the time flowers.

The flowers grew to a height of about six feet, their slender stems, like rods of glass, bearing a dozen leaves, the once transparent fronds frosted by the fossilized veins. At the peak of each stem was the time flower, the size of a goblet, the opaque outer petals enclosing the crystal heart. Their diamond brilliance contained a thousand faces, the crystal seeming to drain the air of its light and motion. As the flowers swayed slightly in the evening air, they glowed like flame-tipped spears.

Many of the stems no longer bore flowers, and Axel examined them all carefully, a note of hope now and then crossing his eyes as he searched for any further buds. Finally he selected a large flower on the stem nearest the wall, removed his gloves and with his strong fingers snapped it off.

As he carried the flower back on to the terrace, it began to sparkle and deliquesce, the light trapped within the core at last released. Gradually the crystal dissolved, only the outer petals remaining intact, and the air around Axel became bright and vivid, charged with slanting rays that flared away into the waning sunlight. Strange shifts momentarily transformed the evening, subtly altering its dimensions of time and space. The darkened portico of the house, its patina of age stripped away, loomed with a curious spectral whiteness as if suddenly remembered in a dream.

Raising his head, Axel peered over the wall again. Only the farthest rim of the horizon was lit by the sun, and the great throng, which before had stretched almost a quarter of the way across the plain, had now receded to the horizon, the entire concourse abruptly flung back in a reversal of time, and appeared to be stationary.

The flower in Axel's hand had shrunk to the size of a glass thimble, the petals contracting around the vanishing core. A faint sparkle flickered from the centre and extinguished itself, and Axel felt the flower melt like an ice-cold bead of dew in his hand.

Dusk closed across the house, sweeping its long shadows over the plain, the horizon merging into the sky. The harpsichord was silent, and the time flowers, no longer reflecting its music, stood motionlessly, like an embalmed forest.

For a few minutes Axel looked down at them, counting the flowers which remained, then greeted his wife as she crossed the terrace, her brocade evening dress rustling over the ornamental tiles.

"What a beautiful evening, Axel." She spoke feelingly, as if she were thanking her husband personally for the great ornate shadow across the lawn and the dark brilliant air. Her face was serene and intelligent, her hair, swept back behind her head into a jewelled clasp, touched with silver. She wore her dress low across her breast, revealing a long slender neck and high chin. Axel surveyed her with fond pride. He gave her his arm and together they walked down the steps into the garden.

"One of the longest evenings this summer," Axel confirmed, adding: "I picked a perfect flower, my dear, a jewel. With luck it should last us for several days." A frown touched his brow, and he glanced involuntarily at the wall. "Each time now they seem to come nearer."

His wife smiled at him encouragingly and held his arm more tightly.

Both of them knew that the time garden was dying.

Three evenings later, as he had estimated (though sooner than he secretly hoped), Count Axel plucked another flower from the time garden.

When he first looked over the wall the approaching rabble filled the distant half of the plain, stretching across the horizon in an unbroken mass. He thought he could hear the low, fragmentary sounds of voices carried across the empty air, a sullen murmur punctuated by cries and shouts, but quickly told himself that he had imagined them. Luckily, his wife was at the harpsichord, and the rich contrapuntal patterns of a Bach fugue cascaded lightly across the terrace, masking any other noises.

Between the house and the horizon the plain was divided into four huge swells, the crest of each one clearly visible in the slanting light. Axel had promised himself that he would never count them, but the number was too small to remain unobserved, particularly when it so obviously marked the progress of the advancing army. By now the forward line had passed the first crest and was well on its way to the second; the main bulk of the throng pressed behind it, hiding the crest and the even vaster concourse spreading from the horizon. Looking to left and right of the central body, Axel could see the apparently limitless extent of the army. What had seemed at first to be the central mass was no more than a minor advance guard, one of many similar arms reaching across the plain. The true centre had not yet emerged, but from the rate of extension Axel estimated that when it finally reached the plain it would completely cover every foot of ground.

Axel searched for any large vehicles or machines, but all was amorphous and unco-ordinated as ever. There were no banners or flags, no mascots or pike-bearers. Heads bowed, the multitude pressed on, unaware of the sky.

Suddenly, just before Axel turned away, the forward edge of the throng appeared on top of the second crest, and swarmed down across the plain. What astounded Axel was the incredible distance it had covered while out of sight. The figures were now twice the size, each one clearly within sight.

Quickly, Axel stepped from the terrace, selected a time flower from the garden and tore it from the stem. As it released its compacted light, he returned to the terrace. When the flower had shrunk to a frozen pearl in his palm he looked out at the plain, with relief saw that the army had retreated to the horizon again.

Then he realized that the horizon was much nearer than previously, and that what he assumed to be the horizon was the first crest.

When he joined the Countess on their evening walk he told her nothing of this, but she could see behind his casual unconcern and did what she could to dispel his worry.

Walking down the steps, she pointed to the time garden. "What a wonderful display, Axel. There are so many flowers still."

Axel nodded, smiling to himself at his wife's attempt to reassure him. Her use of "still" had revealed her own unconscious anticipation of the end. In fact a mere dozen flowers remained of the many hundred that had grown in the garden, and several of these were little more than buds—only three or four were fully grown. As they walked down to the lake, the Countess's dress rustling across the cool turf, he tried to decide whether to pick the larger flowers first or leave them to the end. Strictly, it would be better to give the smaller flowers additional time to grow and mature, and this advantage would be lost if he retained the larger flowers to the end, as he wished to do, for the final repulse. However, he realized that it mattered little either way; the garden would soon die and the smaller flowers required far longer than he could give them to accumulate their compressed cores of time. During his entire lifetime he had failed to notice a single evidence of growth among the flowers. The larger blooms had always been mature, and none of the buds had shown the slightest development.

Crossing the lake, he and his wife looked down at their reflections in the still black water. Shielded by the pavilion on one side and the high garden wall on the other, the villa in the distance, Axel felt composed and secure, the plain with its encroaching multitude a nightmare from which he had safely awakened. He put one arm around his wife's smooth waist and pressed her affectionately to his shoulder, realizing that he had not embraced her for several years, though their lives together had been timeless and he could remember as if yesterday when he first brought her to live in the villa.

"Axel," his wife asked with sudden seriousness, "before the garden dies . . . may I pick the last flower?"

Understanding her request, he nodded slowly.

One by one over the succeeding evenings, he picked the remaining flowers, leaving a single small bud which grew just below the terrace for his wife. He took the flowers at random, refusing to count or ration them, plucking two or three of the smaller buds at the same time when necessary. The approaching horde had now reached the second and third crests, a vast concourse of labouring humanity that blotted out the horizon. From the terrace Axel could see clearly the shuffling, straining ranks moving down into the hollow towards the final crest, and occasionally the sounds of their voices carried across to him, interspersed with cries of anger and the cracking of whips. The wooden carts lurched from side to side on tilting wheels, their drivers struggling to control them. As far as Axel could tell, not a single member of the throng was aware of its overall direction. Rather, each one blindly moved forward across the ground directly below the heels of the person in front of him, and the only unity was that of the cumulative compass. Pointlessly, Axel hoped that the true centre, far below the horizon, might be moving in a different direction, and that gradually the multitude would alter course, swing away from the villa and recede from the plain like a turning tide.

On the last evening but one, as he plucked the time flower, the forward edge of the rabble had reached the third crest, and was swarming past it. While he waited for the Countess, Axel looked at the two flowers left, both small buds which would carry them back through only a few minutes of the next evening. The glass stems of the dead flowers reared up stiffly into the air, but the whole garden had lost its bloom.

* * *

Axel passed the next morning quietly in his library, sealing the rarer of his manuscripts into the glass-topped cases between the galleries. He walked slowly down the portrait corridor, polishing each of the pictures carefully, then tidied his desk and locked the door behind him. During the afternoon he busied himself in the drawing-rooms, unobtrusively assisting his wife as she cleaned their ornaments and straightened the vases and busts.

By evening, as the sun fell behind the house, they were both tired and dusty, and neither had spoken to the other all day. When his wife moved towards the music-room, Axel called her back.

"Tonight we'll pick the flowers together, my dear," he said to her evenly. "One for each of us."

He peered only briefly over the wall. They could hear, less than half a mile away, the great dull roar of the ragged army, the ring of iron and lash, pressing on towards the house.

Quickly, Axel plucked his flower, a bud no bigger than a sapphire. As it flickered softly, the tumult outside momentarily receded, then began to gather again.

Shutting his ears to the clamour, Axel looked around at the villa, counting the six columns in the portico, then gazed out across the lawn at the silver disc of the lake, its bowl reflecting the last evening light, and at the shadows moving between the tall trees, lengthening across the crisp turf. He lingered over the bridge where he and his wife had stood arm in arm for so many summers—

"Axel!"

The tumult outside roared into the air, a thousand voices bellowed only twenty or thirty yards away. A stone flew over the wall and landed among the time flowers, snapping several of the brittle stems. The Countess ran towards him as a further barrage rattled along the wall. Then a heavy tile whirled through the air over their heads and crashed into one of the conservatory windows.

"Axel!" He put his arms around her, straightening his silk cravat when her shoulder brushed it between his lapels.

"Quickly, my dear, the last flower!" He led her down the steps and through the garden. Taking the stem between her jewelled fingers, she snapped it cleanly, then cradled it within her palms.

For a moment the tumult lessened slightly and Axel collected himself. In the vivid light sparkling from the flower he saw his wife's white, frightened eyes. "Hold it as long as you can, my dear, until the last grain dies."

Together they stood on the terrace, the Countess clasping the brilliant dying jewel, the air closing in upon them as the voices outside mounted again. The mob was battering at the heavy iron gates, and the whole villa shook with the impact.

While the final glimmer of light sped away, the Countess raised her palms to the air, as if releasing an invisible bird, then in a final access of courage put her hands in her husband's, her smile as radiant as the vanished flower.

"Oh, Axel!" she cried.

Like a sword, the darkness swooped down across them.

Heaving and swearing, the outer edges of the mob reached the knee-high remains of the wall enclosing the ruined estate, hauled their carts over it and along the dry ruts of what once had been an ornate drive. The ruin, formerly a spacious villa, barely interrupted the ceaseless tide of humanity. The lake was empty, fallen trees rotting at its bottom, an old bridge rusting into it. Weeds flourished among the long grass in the lawn, overrunning the ornamental pathways and carved stone screens.

Much of the terrace had crumbled, and the main section of the mob cut straight across the lawn, by-passing the gutted villa, but one or two of the more curious climbed up and searched among the shell. The doors had rotted from their hinges and the floors had fallen through. In the music-room an ancient harpsichord had been chopped into firewood, but a few keys still lay among the dust. All the books had been toppled from the shelves in the library, the canvases had been slashed, and gilt frames littered the floor.

As the main body of the mob reached the house, it began to cross the wall at all points along its length. Jostled together, the people stumbled into the dry lake, swarmed over the terrace and pressed through the house towards the open doors on the north side.

One area alone withstood the endless wave. Just below the

terrace, between the wrecked balcony and the wall, was a dense, six-foot-high growth of heavy thorn-bushes. The barbed foliage formed an impenetrable mass, and the people passing stepped around it carefully, noticing the belladonna entwined among the branches. Most of them were too busy finding their footing among the upturned flagstones to look up into the centre of the thorn bushes, where two stone statues stood side by side, gazing out over the grounds from their protected vantage point. The larger of the figures was the effigy of a bearded man in a high-collared jacket, a cane under one arm. Beside him was a woman in an elaborate full-skirted dress, her slim, serene face unmarked by the wind and rain. In her left hand she lightly clasped a single rose the delicately formed petals so thin as to be almost transparent.

As the sun died away behind the house a single ray of light glanced through a shattered cornice and struck the rose, reflected off the whorl of petals on to the statues, lighting up the grey stone so that for a fleeting moment it was indistinguishable from the long-vanished flesh of the statues' originals.

Acknowledgments

While working for a different publisher from the one I'm at now, I organized a small reissue of Clifton Fadiman's two classic anthologies from the late 1950s and early 1960s, *Fantasia Mathematica* and *The Mathematical Magpie.* The revival of those books, which provided the template for *Imaginary Numbers,* was the first step of a more ambitious plan that included a new volume to be edited by Martin Gardner.

Unfortunately, Martin wasn't thrilled with the idea; he was busy with other projects he found more compelling, despite my entreaties. I used to call him up every couple of months with another idea for the anthology, until finally he said, "Look, Bill, if you're so keen to get this book done, why don't you edit the damn thing yourself?" Until then, the possibility had never occurred to me.

Martin and his wife, Charlotte, graciously put up with me for two days in 1996 while I ransacked his large and exquisitely well-organized library for material. He directed me to several pieces I would never otherwise have found.

Barry Cipra, a mathematician and longtime friend, turned out to have a wonderful library of mathematically based prose and poetry, of which this collection barely scratches the surface. Both he and Martin Gardner could easily have been credited as co-editors of this book, if not for their charming modesty and my own reluctance to share royalties. Gary Cornell, Arthur Goldwag,

April Reynolds, and Elizabeth Ziemska all made helpful and important suggestions.

Randy Pink cleared the permissions with unfailing skill, knowledge, and diligence; it's impossible to imagine that anyone could have been better at this surprisingly complex task. The various agents, publishers, executors, and authors with whom Randy corresponded were both quick and generous in granting rights to reprint the works they controlled.

My agent, Susan Rabiner, and my editor at Wiley, Emily Loose, have both supported this project with enthusiasm and professionalism—the ideal combination of qualities.

Finally, my family—my wife, Candace, and my two sons, Paul and Craig—make this and everything else I do worth attempting.

My heartfelt thanks to all.

A note on future collections: Readers who find this anthology frustratingly incomplete may send their comments to me, care of the publisher, at the address given on the copyright page, or directly to me at wfrucht@snet.net. I will gratefully consider all suggestions of material for a sequel (previously published work only, please). Older pieces, translations, and works by writers not known for an interest in mathematics are especially desirable.

Permission Acknowledgments

"The Form of Space" is from *Cosmicomics* by Italo Calvino, copyright © by Giulio Einaudi editore s.p.a., Torino, English translation by William Weaver copyright © 1968 and renewed 1996 by Harcourt Brace & Company and Jonathan Cape Limited, reprinted by permission of Harcourt Brace & Company and Jonathan Cape Ltd.

"A New Golden Age," by Rudy Rucker, originally appeared in the Randolph-Macon Women's College Alumnae Bulletin, copyright © 1981 by the Alumnae Association of Randolph-Macon Women's College, and is reprinted by the Association's permission.

"How Kazir Won His Wife" is from *Satan, Cantor and Infinity* by Raymond Smullyan, copyright © 1992 by Raymond Smullyan. Reprinted by permission of Alfred A. Knopf, Inc.

"11 May 1905" is from *Einstein's Dreams* by Alan Lightman, copyright © 1992 by Alan Lightman. Reprinted by permission of Pantheon Books, a division of Random House, Inc.

"Why Does Disorder Increase in the Same Direction of Time as That in Which the Universe Expands?" by Roald Hoffman, copyright © Roald Hoffman 1994, was first published in *The Madison Review* 16(2), 45 (1994), and is reprinted by permission of the author.

"The Golden Man" by Philip K. Dick is reprinted by permission of the author and the author's agents, Scovil Chichak Galen Literary Agency, Inc.

"The Morphology of the *Kirkham* Wreck," by Hilbert Schenck, copyright © 1978 by Hilbert Schenck, first appeared in *The Magazine of*

Fantasy and Science Fiction and is reprinted by permission of the author and the author's agent, the Virginia Kidd Agency, Inc.

"Resolution of the Paradox: A Philosophical Puppet Play," by Abner Shimony, copyright © 1970 by Abner Shimony, is from Wesley C. Salmon, ed., *Zeno's Paradoxes* (Indianpolis and New York: The Bobbs-Merrill Co., 1970), pp. 1–3. Reprinted by permission of the author and editor.

"Parallelism," by Piet Hein © Grook, page 113, is reproduced with kind permission from Piet Hein a/s, DK-5500 Middelfart.

"Prelude . . ." is from *Gödel, Escher, Bach: An Eternal Golden Braid* by Douglas Hofstadter, copyright © 1979 by Basic Books, Inc. Reprinted by permission of Basic Books, a member of Perseus Books, L.L.C.

"The Third Sally, or The Dragons of Probability" is from *The Cyberiad* by Stanislaw Lem, copyright © 1974 by Seabury Press; © 1985 by Harcourt Brace & Company. Reprinted by permission of the publisher.

"On Fiddib Har" is from *The Planiverse: Computer Contact with a Two-Dimensional World* by Alexander K. Dewdney, copyright © 1984 by Alexander K. Dewdney. Reprinted by permission of the author.

"The Church of the Fourth Dimension" is from *The Unexpected Hanging* by Martin Gardner, copyright © 1969 by Martin Gardner. Reprinted by permission of the author.

"The Extraordinary Hotel, or The Thousand and First Journey of Ion the Quiet," by Stanislaw Lem, originally appeared in *Stories About Sets*, by N. Ya. Vilenkin, English translation by Scripta Technica, copyright © 1968 by Academic Press, and is reprinted by permission of the publisher.

The poems by Racter originally appeared in *The Policeman's Beard is Half-Constructed* by Racter (a computer program written by W. Chamberlain and T. Etter), copyright © 1984, and are reprinted by permission of William Chamberlain and the illustrator, Joan Hall.

"Burning Chrome," by William Gibson, copyright © 1982 by Omni Publications International, Ltd., is reprinted by permission of the author.

"Gonna Roll the Bones," by Fritz Leiber, copyright © 1967 by Fritz Leiber, originally appeared in *Dangerous Visions,* edited by Harlan Ellison, and is reprinted by permission of the author's agent, Richard Curtis Associates, Inc.